The Future Was Looking Up

How a small Oregon company changed aviation

Philip A. Moylan

ISBN 978-0-359-22239-1

Contents

Introduction

When we travel, my wife, Chris, always asks if the airplane we're boarding has the Head-Up Guidance System (HGS), and if I say no, she gets a worried look on her face. I have to reassure her that airplanes were flying safely before the HGS, but they are just much safer now.

Anyone who is interested in the world of aviation, or has been involved in cockpit, display, or avionics development, may be curious to know how Head-Up Displays (HUD) evolved and eventually became standard on many new air transport and business aircraft. I hope this book will shed some light on the commercial HUD's early development, successes, and eventual acceptance within the entire air transport industry.

I believe that no new commercial aircraft will be designed and developed in the future without HUD, either as basic equipment or a customer option. I am proud of the role I played in helping to make this happen.

Phil Moylan
Dana Point, California
December, 2018

Chapter 1

Frozen Turkey Heads

We were in the middle of lunch. A relaxed and enjoyable lunch. Suddenly, Jim leaned in close and almost whispered to me.

"It's like selling frozen turkey heads."

What? The analogy totally escaped me.

I wasn't at all sure what anybody would do with frozen turkey heads, never mind trying to sell them. I had accepted the invitation to lunch, perhaps for the wrong reason. Jim had mentioned to me that he was a member of the Bellevue Athletic Club, a rather exclusive place not too far from my office. I was anxious to test it out as a potential lunch spot for visiting company personnel, and possibly for customers. Besides, it was a beautiful summer day in Seattle, Washington, with a sky as blue as that song I couldn't remember the name of, and I wanted to get out of the office for a while.

Our conversation was to change aviation!

Jim had called me at my office the day before to invite me to lunch. He was fishing. He'd spoken to a couple of his acquaintances at Boeing to ask if they knew someone who might be a good fit for a marketing position he was trying to fill, and my name had come up twice. I suppose I should have been flattered. They had given Jim my office telephone number, and I had accepted his invitation.

Jim Gooden was Director of Marketing for a small aerospace company in Hillsboro, Oregon, called Flight Dynamics Inc. I had never heard of the company, let alone Hillsboro, and I'd never been to Oregon. Oh, I knew our neighboring state to the south was big in timber and fish, but this Flight Dynamics product was a high technology Head Up Display (HUD) for the commercial air transport marketplace. The only thing I knew about HUDs was what I had learned from some war movies of the time, in which aerial sequences were filmed through the HUD on military fighter jets like the F-14. But I had absolutely no idea why anyone would want to put a HUD on a commercial airliner.

"We've been at this for well over five years now," Jim continued, "and we're getting nowhere fast. I need real help! I need someone with good innovative thinking to get behind this thing. Think you might be interested, Phil?" He was definitely very keen in his pursuit.

As I looked admiringly around the club, I must admit I was more interested in checking the place out, and enjoying the food, than in what Jim was saying. But although I had just met him, I liked Jim. He was older—perhaps 60—quite weathered and persistent, but very personable. And he was focused. Actually, I'd never met anyone quite so focused on one subject as Jim was on this HUD thing of his.

"But Jim, don't you want someone who has experience doing this?" I asked. After all, I knew very little about HUDs or the relevant technology, and I was unsure how exactly I could be of any real assistance to his little company. I also did not have much contact with the airline segment of the aerospace industry, since my current focus was mainly on Boeing.

"No one has experience doing this Phil, no one!" Jim explained. "This is new—very new—and very tough. It's certainly not for the faint of heart."

Geez I thought, if he's trying to persuade me, he's going about it all wrong.

My current employer, Simmonds Precision, had transferred me from Toronto, Canada, to the Seattle area three years before to become their Pacific Northwest Director of Marketing. Simmonds was a US East Coast manufacturing company with divisions involved in aircraft fuel management, flight controls, and engine systems. I had made some good progress with sales up north to manufacturers like Canadair, deHavilland, and Pratt & Whitney Canada. And I had pulled off a huge government product support contract for the CF-18, the new Canadian fighter jet at that time. But now that I was in the US, my primary customer was Boeing. I was feeling confident in my responsibilities, and I had also been lucky at my new role, with a few successes here and there. I supposed that was why Jim had heard about me.

But Jim could tell I wasn't all there that day. "Tell you what," he said excitedly. "I'll set you up for a simulator demonstration and then we can talk some more."

"Wait, what? A simulator? But...but I'm not a pilot," I stammered. "I don't fly."

"Doesn't matter." Jim turned to wave the waiter over for the check. "You'll be flying a Boeing 727 five minutes after you get in."

This got my attention. Five minutes…yeah, right!

Back in my Simmonds office later that afternoon, I cornered our technical support fellow to ask him what he knew about HUD. Basically, he reiterated what I had already seen in some Hollywood movies, but he knew virtually nothing at all about commercial aircraft HUDs or why they might be needed. He had never heard of that. And he did not know anyone technical that I could call for more information. He and my secretary both became concerned with my interest in another company's product, and what they perceived as a possible job interview. I laughed it off—no way, nothing to worry about!

My discussion regarding the sale of frozen turkey heads had taken place on a beautiful Wednesday in August of 1985. I barely thought about it at home that night and didn't even mention it to my wife. But early the very next morning at my office, I got my second call from Jim, this time to tell me he had a simulator session already booked for me the following day at 10 am. Jim was definitely not one to let any grass grow under his feet. I liked that too. He gave me directions to the "M-Cab," Boeing's multi-purpose simulator. It was located at their Renton facility, about 30 miles south of my office, and Jim advised me of the best route. Once again, I reminded him that I was not a pilot and that I had never, ever flown an aircraft of any kind before.

"Don't be late," was all he said, then hung up.

After I got off the phone with Jim, my secretary gave me an alarmed look and asked me why I was going to a Boeing simulator to see this HUD thing. I told her I was just curious, which was completely true. In reality though, Jim had sparked my interest in this small company and their unusual aviation product.

That afternoon, I called some other contacts I had in the aerospace industry to see if I could dig up any more useful HUD information. Again, I ran into a dead end. Some knew a little about military fighter jet HUDs and their role in displaying weapons status to the pilot, but none had ever heard about HUDs for commercial airliners like the Boeing 727.

The next day, I headed south to Renton. After many visits to Boeing on the job with Simmonds, I knew exactly where Jim wanted to meet me. He was already there and gave me a bright, welcoming smile. We signed in at the lobby and I followed him through a maze of cubicles, offices, and hallways until we ended up in a small control room. I spied what definitely looked like a flight simulator through a connecting window. There were three other people in the room. One was seated behind a large computer screen, busily entering data on a keyboard. Jim pulled me over to the second person, who looked quite serious and was preoccupied with some paperwork.

"This is John Desmond," Jim announced. We shook hands and Jim moved me on to the third fellow, who wore a huge smile that extended clear across his face.

"Hi! Good to meet you. I'm Mike Gleason," he almost shouted. "So, you wanna fly our HUD, huh? This is the best thing to happen to aviation since the invention of the jet engine." I was quite amused with his enthusiasm, which went a long way to ease my nervousness.

Captain Mike Gleason was about my age, slim and pleasant-looking, with a comical demeanor. After we shook hands, he began to walk across a small bridge to the simulator door and beckoned me to follow him.

As we crossed the bridge, I tapped him gently on the shoulder.

"Hey Mike, I gotta tell you something. I'm not a pilot, and I've never flown an airplane before…ever." I felt awkward, and hoped Mike understood.

But, like Jim, he ignored my comments.

I still couldn't believe these guys were going to put me into a flight simulator and expect me to fly. This had to be a joke. As I entered and approached the pilot seat, it was hard to miss the HUD. A thick hunk of glass about the size of an 8x10 piece of paper hung in front of the pilot's eye position. And there was a "bump" on the ceiling over the head area. Mike told me to sit down and take a look through the glass. As my eyes focused on a myriad of green symbols and numbers, I was awestruck at how clear everything was, and although I had absolutely no idea what I was looking at, I was very impressed.

As he took the co-pilot seat, Mike told me that the symbology on the display was actually focused at infinity, and he held up his business card right in front of the glass screen to prove his point. I found that I had to refocus my eyes to read his card. Very clever.

He then went through the various numbers displayed around the periphery of the glass, calling out what they represented. They were terms I'd heard before in my aviation career, but I couldn't put them in any useful context: airspeed, barometric altitude, vertical speed, DME, heading. Then he got to the middle of the display and began talking a whole new language. The circle with the canted legs...well, that was "the flat-footed duck," and next to it was the "carrot." The duck was like a gunsight, Mike explained, and because I could see the outside world through the glass, I could put the duck exactly where I wanted to land. If it lay over the grass, well, that was our touchdown point. The carrot—actually, the "caret"—was a display of the aircraft acceleration or deceleration, an indication of its instantaneous energy state. I had no clue why this was in any way relevant. Next, he brought up a smaller circle and told me that was the guidance cue. All I had to do was fly the duck over the cue. That sounded simple enough. Placing his left hand on the throttles, he told me he would look after everything else.

He reached up to an overhead panel and pushed a button.

"OK, Phil," said a grinning Mike, "you're flying!"

Flying? You're kidding! How could I be flying after only two minutes of instruction? And this was a Boeing 727 jetliner. Totally focused on the tracking task at hand, I managed a quick glance down at the sea of round dials on the instrument panel in front of me, some moving, some not. Back to the duck and cue. After only a minute or two of "flying," I found the task to be fairly simple and straightforward. As the ground approached, my nervousness returned, but Mike kept at me. "Follow the cue...follow the cue."

Before I knew it, we were down. A perfect landing, on my very first try.

"Holy crap...that's incredible!" I exclaimed, breathing a visible sigh of relief.

"Now we're gonna do it again," explained Mike, "but this time in CAT3."

“What the hell is CAT3?” I asked.

“Lousy fog,” responded Mike, preoccupied with resetting the simulator. “What pilots call ‘sporty landing conditions’.”

When you live in the Pacific Northwest, moving around in fog gets to be second nature. But for the first time, I wondered how pilots flew aircraft to a safe landing in fog when I could hardly see to drive to the neighborhood 7-Eleven. Mike set me up and once again I was at the controls. But this time, all the outside references—ground, runway, terminal buildings—were gone, and I stared into a sea of gray.

“Is this what the pilot actually sees?” I asked.

“You bet,” said a contented Mike as he monitored what I was doing on a small screen on his side of the cockpit. I hadn’t even noticed it when I climbed into the pilot seat. It had all the same symbols on it but no real world. Mike could tell how well I was tracking the HUD’s guidance cue just by watching this little repeater display.

It was eerie. Staring out into that gray, I realized just how difficult it must be for a pilot in a real airplane really flying in these conditions with real passengers behind him. The term “sporty” suddenly took on a whole new meaning. But I simply flew the duck over the cue as instructed and suddenly, through the murk, I glimpsed runway lights…and then the runway. I was spot on. After landing, as before, right on the centerline and in the runway’s touchdown zone, I became somewhat suspicious.

“Is this a real flight simulator, Mike?” I began to doubt how I could possibly have landed a B727 in heavy fog conditions after only ten minutes “behind the wheel,” without any prior flying experience whatsoever and virtually no instruction.

“Sure is!” Mike replied, really beaming now. “Works exactly the same way in an actual 727 aircraft,” he assured me.

I was incredulous. If someone like me who is not a pilot can do this, how much more could the HUD do for a real pilot with years of training and flying experience?

Mike and I had been in the M-Cab for well over an hour. I found myself doing takeoffs, visual and “black hole” approaches, circling, and an assortment of other typical operations encountered by airline pilots in the real world of flying. Black holes (not the space phenomena) are treacherous nighttime flying scenarios where there are no city or horizon lights to help orient the flight crew. Often, the pilot gets into Pilot-Induced Oscillation, called PIO. Mike explained that without a HUD, this scenario can quickly become a serious challenge.

As we chatted during the resetting of the simulator, I learned a bit about my instructor. Mike had been a DC-10 pilot with Continental Airlines for many years, but there had been a major shakeup at the company, followed by some union issues and a big layoff. He had taken the job with Flight Dynamics nearly a year ago, but was still living in Malibu, California. (Now there's a commute.)

As we exited the simulator across the adjoining bridge, there was Jim, now also beaming. Somehow, he knew.

"Well," he asked, "what do you think?"

Now it was my turn to beam. "I've never seen anything like it. That's really amazing. But why won't it sell?"

Jim explained to me that their HUD was not cheap and that airlines had to justify the purchase of such an expensive piece of equipment. The company had not yet been able to convince any airline of the value of such an investment.

"That's gonna be your job," Mike chimed in from behind. He was grinning from ear to ear and visibly pleased with his demonstration skills.

For the first time, I began to seriously consider the offer Jim had hinted he was ready to make me.

Jim had been right earlier that week over lunch. Five minutes with this HUD was all that was needed to learn to perform some of the most difficult tasks facing a pilot in daily line operations. Flying just couldn't be that easy. There had to be more to it. Especially with a big airliner full of people. As we were wrapping up and getting ready to leave, Jim asked if I could make the trip down to Hillsboro the following week to meet some of the other players, including John Geiger, the company's president. My schedule supported a Thursday visit so I agreed.

This was definitely worth further exploration.

After I returned to my office, I had to again reassure my increasingly worried secretary, although I realized I was not doing such a good job of convincing her.

Early the following week I was on one of my routine visits to Boeing to check on the status of some equipment deliveries from Simmonds. I asked my procurement agent if he knew anything about HUDs. Again, nothing! He suggested that some of their Boeing engineering folks might have more information. I decided to call one of my senior cockpit electronics contacts there to explore the subject. His name was Jack Brit.

Jack kindly made some time for me in his always-busy schedule, and later that morning, I found myself sitting across from him at his desk.

"So Phil, what's up, what's so pressing?" he asked, looking up from his paperwork and leaning back in his chair.

"Well…I was wondering what you thought of HUDs, Jack," I said, watching carefully for his reaction.

"Great! For military fighter aircraft," was his reply. "But not much use to us here. We usually don't put weapons on our 737s."

"What about for flying in foggy weather…like for CAT3?" I decided to show off my newfound vocabulary.

"We developed automatic landing systems many years ago to handle those difficult situations. All of our production aircraft have CAT3 autoland and simply don't need a HUD," came the answer from Jack.

I was devastated. Somehow I had expected, or hoped for, a different reaction.

"What about for "black holes" and such?" I continued to show off.

"Look, Phil," Jack was now becoming visibly annoyed, "Boeing is not remotely interested in HUD for any of our airplanes, and it'll be a cold day in hell when I would endorse putting one on." His voice grew louder. "Did you want anything else?"

If ever I'd heard an obvious invitation to exit, that was it. I thanked him and left, feeling very down. Back in my office in Bellevue later that day, I thought a lot about his reaction and what it could possibly mean. How could something so amazing be of so little interest to one of Boeing's specialists? Did he know how good this HUD was? Or did he know something I didn't? Why wasn't he even a little bit curious or interested?

My Simmonds technical support fellow, and especially my secretary, seemed quite relieved, and laughed heartily when I told them the story over lunch the next day at one of our favorite restaurants. However, I failed to see the humor in it. For my pending trip to Oregon later that week, I decided to say I needed a couple days off so as not to alert them that I was going to a continuing job interview.

On the four-hour drive to Hillsboro a few days later, Jack's words and reaction tumbled around in my mind. He had reacted almost violently. Certainly dismissively. What the hell was I not getting here?

As I drove along, I struggled with the opposing feelings I had: the elation of landing a B727 jetliner after just a few minutes of instruction and without any prior training or flying experience whatsoever, against Jack's dismal prediction about the future of HUD at Boeing. I knew enough about the aerospace business to realize that, without a Boeing endorsement, the HUD would eventually die on the vine.

But the beautiful scenery on my drive south distracted me from thoughts of business. I had moved to Everett, Washington, over three years ago, and vividly recalled a conversation I'd had with one of the Boeing buyers over lunch a few weeks after I'd moved to the Pacific Northwest. He warned me that if I stayed in this part of the country for two years, I would never leave it. As I passed the colorful fields of flowers and lush greenery extending all the way to the snow-capped mountains on the horizon, I had to agree with him. I loved it, and I knew I would be here a long time.

Hillsboro was a small rural community about a fifteen-minute drive southwest of the city of Portland. Using the hand-drawn map that Jim had given me as we left the simulator, I finally found my way to the company's headquarters. It was a small, nondescript building in the midst of a huge complex of similar buildings with varying corporate names and logos above their front doors. After finding the one labeled "Flight Dynamics Inc," I entered and waited patiently in the receptionist area while they tracked Jim down. Finally, he burst through the adjoining door and greeted me with a smile.

"I see you made it OK!" He was slightly out of breath, and seemed genuinely glad to see me.

First was a brief tour of their facility. It reminded me of many of the labs we'd had in the aeronautical engineering department at university back in Canada: oscilloscopes, computers, and assorted pieces of what I presumed were test equipment. On the tour, Jim introduced me to some of the folks manning the equipment, and others we ran into in the hallways: Bob Wood, Doug Ford, Ken Zimmerman, Norman Jee. I wasn't sure I'd remember their names, but they all looked contented, and really excited about what they were doing. Their excitement made me forget Boeing's negative comments for the moment.

Next up was a meeting with John Desmond, whom I had met briefly at the M-Cab. Jim led me into John's office and we sat down. John was a little bit older than me and was Executive VP for the company. He had quite the office, with a huge window and expensive-looking furniture. There was even an executive bathroom attached. After we got comfortable, he asked many questions about my current job at Simmonds, my responsibilities, and my background, and seemed content with all my answers.

"So how would you go about selling this HUD of ours?" John finally asked.

Trying hard not to laugh, I smiled as I was reminded of Jim's comments regarding frozen turkey heads. I considered telling him about my awful conversation with Boeing earlier in the week, but decided against it for the moment.

"Well, first we'd have to find an airline customer with a significant problem that the HUD fixes," I responded almost flippantly, although flippancy was not what I wanted to convey. "And then we'd have to convince their folks to spend the money to fix it."

My response seemed to satisfy my interviewer, and I could tell that a smiling Jim was pleased with the interaction.

After we left John, Jim led me back to his office and we sat down. I finally summoned the courage to tell him about my depressing discussion with Boeing earlier in the week. Jim just smiled, the last reaction I expected. He'd obviously been through that before.

"Turkey heads, remember?" he almost chuckled.

"So," I pressed him, "have you sold any at all?"

"Nope," was Jim's reply, "not one."

Holy crap! What was I getting myself into?

Jim explained that the B727 was an older aircraft and was not equipped with the autoland that Jack Brit had mentioned, although it was in common use on many newer aircraft. So our HUD was really offering a significant improvement in operational capability to that older 727 aircraft.

He went on to explain to me that the year before, Flight Dynamics had approached Boeing Technical Services to ask about letting them install their HUD in the M-Cab. The two companies had come to an agreement and the HUD, as I had seen, was finally installed. The installation served two purposes: it allowed Flight Dynamics to demonstrate the HUD to potential airline customers, and they hoped it would entice curious Boeing engineering or flight test folks to take a look at it.

Next on my Hillsboro agenda was lunch with "le grand fromage"—the big cheese—John Geiger, the company's president. As Jim drove me over to Maxi's, a little restaurant inside the Hillsboro airport a few miles away, he reassured me that things were going well. And he could tell I was really getting hooked. Having worked for a small company in the '70s before joining Simmonds to run their Canadian office in Toronto, I remembered how much fun it could be.

We parked, entered the restaurant, and arrived at the table just as John Desmond and John Geiger were sitting down. After initial introductions, I studied the president intently. He almost had that banker look: well-dressed, well-educated, serious. I could tell he was intent on finding out more about me. I caught him looking down at my cowboy boots more than once and thought I probably should have been more careful in my clothing selection for a job interview. But the discussion went well and both executives appeared pleased with my interest in their HUD.

"I suppose you would be taking some of the Boeing folks to lunch up there in Seattle?" John Geiger asked.

"Well, yeah...I guess so," I responded. "Why?"

"Well, things are pretty tight for us just now, so maybe you can do Pizza Hut for the first little while," came the answer.

The first error message flashed somewhere on my hard drive. If things were that tight, did I really want to dump my current job for this? After all, I had it pretty nice in Bellevue, with a beautiful, well-equipped office on a small man-made lake, looking out into a little grove of evergreens. I had a great secretary and a knowledgeable technical support assistant, and the three of us got along very well. The company's products were quite familiar to me, since I had run their Canadian office for a couple of years before being transferred to the US. So I felt very comfortable discussing even highly technical issues with the Boeing engineers and buyers. I had a brand-new company car, and my expense account was practically unlimited because of the customer I was doing business with. My bosses were four thousand miles away and left their offices by 2:00 pm my time.

I would be giving up a lot, I thought.

Driving back from the airport, Jim appeared quite pleased with the way things had gone, and my answers to the president's many questions. Now, back in his office, he asked me to wait a while because he had to attend a "huddle" with the executive management—I supposed about me. Lying around Jim's office was an assortment of reprints of articles about the HUD that had appeared in trade magazines familiar to me. As I read through them, it seemed to me that the press was as amazed as I was about the technology, and again Jack Brit's comments returned to haunt me.

Jim was gone about twenty minutes, but finally returned with a contented grin on his face.

"They want to hire you," he blurted out. "When can you start?"

Perhaps it was the challenge of the unknown, or my elated feelings after flying the HUD, but despite the negative comments I'd heard at Boeing, I suddenly decided I wanted to do this. Oh, there was still a lot more to do—salary negotiations, discussions about relocating—but I told Jim I would give four weeks' notice to Simmonds the following week. Jim explained that there was no rush to move to Hillsboro—I could stay in Everett for the time being, since there would be a lot of work to do at Boeing.

That turned out to be an understatement!

Meanwhile, things were looking up…so to speak.

The HUD flight symbols seen through the glass screen, and focused at infinity

(l) Mike Gleason at home with his family in Southern California; (r) Jim Gooden

Chapter 2

Meeting the A-Team

Like everything else about this whole experience, my first day on the job with Flight Dynamics was a bit weird.

Desmond's secretary, Sue Shubert, called me on a Friday in early October, just before I was to start with Flight Dynamics. I had explained to my disappointed and distraught Simmonds team that I was leaving to chase the HUD. They had tried desperately to change my mind. My supervisors back East were not too thrilled with my news either. But, with my help, the company had hired a replacement for me and I had spent time getting him up to speed on the products and the status of the Boeing programs we were involved with. Even so, my team was upset about my departure, and we'd had a sad goodbye lunch at our favorite local restaurant.

"Phil, we don't want you to come down here to Hillsboro on Monday," Sue announced. "Instead, John wants you to fly to New Orleans."

"New Orleans! Why?" I asked.

There was an NBAA, or National Business Aircraft Association, show going on there, and the company was exhibiting the HUD for the very first time. My airline tickets would be waiting at Doug Fox Travel near the Seattle airport, and she had made car and hotel reservations. I'd never been to that part of the country before, so I thought it would be interesting. I knew that the NBAA show was one of the biggest and most popular aviation events in the US, but I had never actually attended one.

My trip to New Orleans was uneventful. I arrived, picked up my rental car, drove to the hotel, and checked in. I decided to walk the one mile to the Convention Center, enjoying my first sights, smells, and sounds of Bourbon Street and the French Quarter as I strolled. After registering for the show, I wandered into the huge facility and finally found Jim and Mike setting up the HUD on one of the many stands being erected inside the building. Flight Dynamics had a small corner of the Pentastar Aviation booth. Jim, smiling as always, came over to shake my hand and congratulate me. He presented me with a little cardboard box which contained a bunch of my new business cards. Mike was his usual funny self, laughing and joking about having a HUD-ache. Then there was Gil McCutchen, busily connecting an octopus of wires to a computer housed in a small cabinet beneath the HUD. He was our technical support for the show. Jim introduced us, Gil stood up, and we shook hands.

Over a great Cajun dinner and drinks that evening in the French Quarter, I found myself fitting in well with the team. Mike, Gil, and Jim talked a lot about the history of the little startup company, founded in 1977 by Bob Carter, a Hillsboro businessman. The HUD was initially developed with private capital. Carter invested a windfall that he'd made while he was working in sales for Floating Point Systems, a microprocessor company located in Beaverton, Oregon, that had done exceedingly well.

Flight Dynamics went public in 1981. In 1983, with a prototype HUD system developed, serious work began on the FAA certification of the system for CAT3 operations on a Boeing 727. They had selected this aircraft since it was the most popular type flying with the airlines at that time. Certification was completed in April 1985, a few months before my "turkey head" lunch with Jim in Bellevue. It had been a long and difficult journey...and very expensive. Mike and Gil were confident that with the HUD finally certified, the airline industry would beat a path to our door to sign up.

Jim wasn't so sure. And after my dreadful Boeing meeting, neither was I!

Over breakfast at the hotel the next morning, we babbled and laughed about what to expect at our first aviation show. I felt it could be a good indicator of what lay ahead for us, for our company, and particularly for me. I also hoped to learn a little more about why I had gotten such a negative reaction about HUD from Boeing. Obviously, I must be missing something critical.

The event was an eye-opening experience for me. This was very obviously the segment of our industry that had lots and lots of money. Multi-million dollar "bizjets" of every size and shape were strewn about the hall like so many new automobiles in a car dealer's showroom. Vendors of luxurious leather seats, deep pile carpeting, airborne telephones and televisions, and aviation electronics—avionics, as they are called—of every type imaginable littered the maze of aisles in the huge convention center. It was impressive!

Gil and Mike turned everything on in our corner of the booth to make sure it was operating correctly. Attendee interest in our HUD was instantaneous. Within ten or fifteen minutes of the official opening of the show on that first morning, a small crowd had lined up around our little booth, wanting to look through the glass, asking questions about the symbology and how the system worked, and scrambling for literature on the product.

I was amazed and elated!

Surely this was going to be easier than selling frozen turkey heads. This was real interest from real customers who had money, and they were asking all the right questions: how much, for what airplane, when can I get one? By the end of that first day, my small business card holder was bulging at the seams with potential customers from what looked like the Fortune 500 companies of America: Xerox, Corning, IBM, Shell Oil, GE, Seagrams. The list was as long as it was impressive. I had written little notes to myself on the backs of their cards. This one wanted outline drawings of the HUD, that one an official quote for a Canadair CL600, another for a Dassault F900. Our team was really buoyed up after such a great first day. Many of our initial visitors, chief pilots for their companies, returned later in the day with others, and sometimes with senior decision makers in their organizations, to show them the HUD and introduce them to us. We even had a few technical visitors from some of the bizjet manufacturers like Dassault and Cessna. Over a great N'Orleans dinner that evening, we laughed heartily about our incredibly successful first day and some of the positive comments we had received from potential customers.

The ensuing two days were very similar, giving me more hope than ever that this was going to be much easier than I had originally thought. And I learned virtually nothing to explain the earlier Boeing reaction.

Although he was also excited about the response to our first show, Jim calmly explained to me that we would need to work closely with the Original Equipment Manufacturers, or OEMs, of these bizjets to get our system installed and integrated on their aircraft and, of course, certified. It was not going to be quite as easy as it looked. Still, I felt that the significant interest we had witnessed at the show could be a huge driver for these OEMs. Strong customer desire always worked in your favor with the aircraft manufacturer.

The next Monday morning I found myself once again on the beautiful drive down to Hillsboro, but this time as an official employee of Flight Dynamics. After arriving, I met with Jim and we once again discussed the successful NBAA show and all the interest in our HUD. Then I was shown to my new office, which looked more like a broom closet. I had a small corner of a bigger room and the furniture had definitely seen better days. But I got to work unpacking my things.

When I was finished, I decided I needed some real HUD education.

After managing to locate Bob Wood, our optics expert, I leveled a barrage of questions at him. How does it work? Why are the symbols green? How heavy is it? Why is the glass so thick and why are the symbols focused at infinity? The poor guy! But Bob, like the others I'd met so far, was enthusiastic, not just about the HUD, but also about my coming onboard, and he wanted to teach me whatever I needed to know to help talk intelligently to potential airline customers. Taking his time, he quickly gauged the level of my engineering understanding and gave me the "HUD-101" technical course. The bump I had noticed over the pilot's head, that was the overhead projector unit. It contained the optics magic—a very complex set of lenses as well as the electronics to project the symbols. The thick hunk of glass in front of the pilot, well that was called the combiner because it combined the now-familiar little green symbols generated by the overhead unit with the real world as seen by the pilot. The combiner, therefore, had to be reflective and yet transparent at the same time—no easy feat.

Then Bob waded into the "holographic sandwich" and I lost him. Were we talking about the HUD or lunch?

Patiently, Bob explained that the flat hunk of glass—the combiner—was actually two pieces of glass sandwiched together with a holographic film in the middle. The film was then exposed with a laser to make it reflective to the green color symbols projected by the overhead unit. OK! Now things were becoming clear…sort of.

I liked Bob Wood immediately. He knew what he was talking about and instilled confidence. I knew I was going to need a lot of that later at Boeing. Bob had been one of the original employees at the little company. He never tired of explaining his brainchild—the entire optics design was his.

The company founder, Bob Carter, had read about the many safety benefits of military fighter HUDs in an aviation magazine. He wanted to adapt that same technology to the commercial aviation marketplace and had hired Bob Wood to work on the HUD optics portion of the project.

Why had Carter chosen Oregon, and why, for God's sake, Hillsboro? Wood explained that Carter owned a couple of aircraft at the Hillsboro airport and wanted to work nearby. Office space was also relatively inexpensive in that area. I assumed that Carter had a home and family there too.

Next on my "hit list" was Doug Ford. He was an outgoing and personable fellow and, as with Bob, I felt immediately relaxed around him, and was impressed with his expertise. He told me he was responsible for the flight controls, whatever that was!

It turned out that Doug's claim to fame was that little guidance cue that allowed me to land so effortlessly in CAT3, right where I was supposed to be. He explained that the cue had to "know" what was going on with the airplane. Its job was to track the electronic runway guidance signals provided from the ground equipment and offer corrections to the pilot if he was deviating. Then, if the aircraft was going faster than usual, or slower, the cue had to adjust for landing flare—a necessary, gentle nose-up move to ease the landing impact. Especially for CAT3 conditions, the cue had to ensure that the plane was spot on, every single time.

The cue was, therefore, one of the trickiest parts of the HUD. Doug got into control laws and all sorts of heavy-duty technical stuff, which had my mind reeling in about five minutes. Although not a pilot, he was extremely knowledgeable about flying and, after seeing some of my confusion, was able to explain all the symbols on the HUD and exactly what they meant in layman's terms, something I really appreciated. I laughed heartily as he referred to the pilot as "the meat servo" and the passengers as "the pink fleshy things in the back."

As we chatted, I knew Doug and I would get along well, and I learned a bit about him along the way. Before coming to Flight Dynamics, he had worked for Boeing Aerospace up in Kent, Washington, where he had developed his flight control credentials working on the autopilot for an air-launched exo-atmospheric missile. He had begun to worry about his contribution to starting World War III and decided he wanted to work on commercial projects. No more "blue coats," he said.

That's when the company found Doug.

Glancing down at my list, I noted the next name: Ken Zimmerman. I found his office and asked him what he did. Ken was quiet, but like Bob and Doug, he knew his stuff. His area of expertise was software. I still had nightmares about software from my Fortran IV programming days at university, and I knew that a simple misplaced period or comma would screw the whole thing up. So my hat went off to Ken. The HUD was software-intensive, so he had a big piece of this puzzle. Like the other two, he was genuinely pleased that I had accepted the job and was eager to assist me. Ken had been hired from Rockwell International, where he had worked on the Space Shuttle's HUD software. (Now there's a credential!) Ken had to define the information and symbology that would be sent to the HUD's combiner , and then develop software that would send it. Patiently and gently, he led me through the logic behind the system, and as he spoke, I finally began to really comprehend what an incredible marvel the HUD was, and how much there was behind my experience of "flying" to a CAT3—or, more accurately, a CATIIIa—landing in the simulator.

He went on to explain that he and Doug had to work together as a team, and they had spent nearly a whole year up at the M-cab simulator trying to characterize how a generic pilot would respond to changes in the aircraft's state. In other words, how quickly can a wide range of possible pilots be expected to react to changes in guidance? Then they had to fine-tune the HUD guidance cue algorithms to provide accurate corrections that were consistent with a pilot's anticipated response to any commands. A real challenge!

With a prototype HUD system now developed, the company had managed to find a B727 operator who was willing to lease his aircraft and allow the team to modify the cockpit to install the HUD for the required certification flight testing. The installation had been done in Tucson and took over a month. Desmond did not want the entire brain trust of the company on the first test flight, just in case, but that first flight went spectacularly well and literally no issues arose.

Ken and Doug both mentioned a real visionary at the FAA (I know, an anomaly), Berk Greene, whom I was to meet many moons later. At the time Flight Dynamics was developing the HUD there were no FAA regulations governing the certification of such a device, so Berk adapted and applied the regulations for the CATIIIa autoland to support the HUD objectives and certification efforts. After flying the HUD in the simulator with Mike Gleason, and later on the actual aircraft out of Tucson, Berk became a huge fan and never gave up, even in the face of some strong objections at the FAA. He was absolutely key to the final FAA certification of the HUD for CATIIIa manually flown operations, which was also key to any potential sale of the system to an airline. His dedication was incredible.

Last on my list was Norman Jee. He was our mechanical engineering genius, and was involved in developing the packaging of the overhead projector unit and the combiner mechanism. He had worked previously for Kaiser Electronics, where he was involved in designing HUDs for military fighters. One of the many challenges he faced on the design of a HUD for commercial airlines was that the combiner had to be able to stow easily out of the way to allow a pilot to get in and out of the seat in the cockpit without banging his head. This was never a requirement in a military fighter aircraft, where the pilot just drops in from above.

It also had to incorporate a break-away feature—that is, the combiner had to move out of the way on its own in the event of an emergency landing, so that a pilot thrown forward by the hard landing would not bash into the combiner glass. This was much later called HIC, or Head Injury Criteria. Norman's design allowed the same force that pushed the pilot forward to also swing the combiner out of the way under its own weight. Again, I was thoroughly impressed.

There were many others to meet in Hillsboro over the ensuing weeks as I got my HUD legs, and all of them amazed me. I could not remember being this impressed with co-workers on any of my previous jobs.

This was truly an A-Team!

Each team member was just as anxious to learn about me, my background, and how I intended to sell "their baby." I began to wonder if I could live up to their expectations, and to the high standard they had established for membership in such an elite club. The products I was used to selling were trivial by comparison to the HUD, and I hoped I could do the job that Jim had hired me for.

As we spent time together over meetings, coffee breaks, lunches, beers, and hallway get-togethers, I also came to the realization that Flight Dynamics was now actually owned by Pacific Telecom, a telecommunications corporation in Oregon.

And there was a big problem.

Our parent company was getting very anxious for an airline fleet HUD sale to justify their acquisition and continued support of the company and our frozen turkey head campaign.

We would definitely need to get a move on.

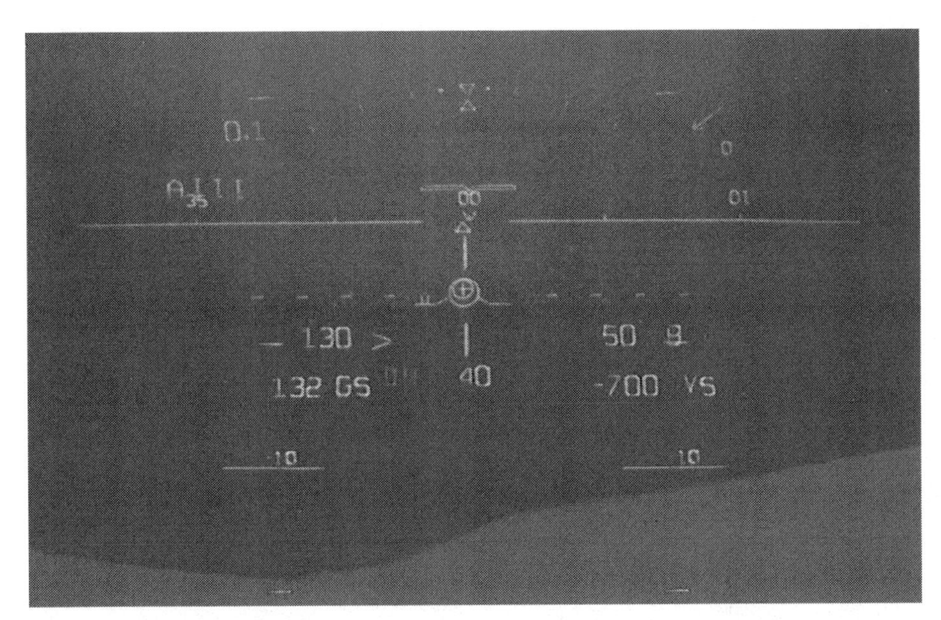

HUD symbology, including the flat-footed duck, the caret, and the guidance cue

John Desmond, Berk Greene (FAA), and Mike Gleason discussing HUD certification

Chapter 3

The Bruce Caulkins Memorial Sandwich

It was time.

Time to leave the nest.

Time to spread my newfound wings!

I had been with Flight Dynamics for over four weeks now, and with all my new education, and trial runs with the A-Team, I was feeling quite confident that I could carry on a reasonable HUD conversation with a potential customer.

Often, members of the team would gang up on me and begin asking questions about the HUD: how it worked, and its many benefits. These were questions that I could expect to encounter out in the real world. Then they would correct any wrong or misleading answers. They also provided useful anecdotes and comments from key industry folks to help explain our product and its value. I knew I could count on any of the A-Team for assistance if and when I ran into a problem. They exuded so much confidence that it was literally contagious.

Jim was in his office. I knocked on his open door and, as always, he beamed and invited me in.

"So, how do you feel about the whole thing now?" he asked, leaning back in his chair.

I made myself comfortable in the seat across from his desk. "I think I'm finally ready!"

Even though I'd said the words, I wasn't sure that Jim believed them.

Hell, I wasn't sure I believed them!

"So, what's your initial target?" he inquired, as he stared at me with that quizzical look of his. I was getting used to it.

I had spent a great deal of time thinking about this one. I knew the question would come up sooner or later, but I had expected it from John Desmond. So I had developed what I thought was a good argument to convince John. And I had run it past Mike Gleason, who was a great help in formulating my justification for the target.

"Alaska Airlines," I blurted out.

"Hmm..." Jim seemed doubtful. "We talked to them already, and they were not interested." I suddenly felt that old nagging doubt I'd had after leaving Jack Brit's office at Boeing that day many weeks earlier. After all my recent efforts, could I still be missing something?

"But Jim," I began my case, "if we can get them as a customer, it proves we don't need initial interest to determine a potential customer. It proves we can go out and get the whole world. We just have to develop a convincing story to help their decision process."

"But why Alaska, Phil?"

"Well, I figure it's like this," I continued, undaunted. "First, they have fog. Bad fog. Almost everywhere in their network. Their hub, Seattle, is the third worst city in the US for fog. And then there's Portland, San Francisco, Sacramento, Oakland, Anchorage…" I had done my homework by checking with the National Oceanographic and Atmospheric Administration (NOAA), and they had provided me with a list of the forty cities in the US with the highest incidence of fog. Alaska Airlines had a lot of them in their route network. Sacramento, for example, was the worst city in the continental US for fog.

"And, with headquarters in Seattle, they are close enough for our folks to drive there," I continued. "This will save us on flights and hotels for staff that need to visit the customer once we sign them up." I smiled as I remembered John Geiger's comment at our first lunch about taking customers to Pizza Hut to save money.

Jim was now smiling at my confidence.

"And they have 727s," I continued, "twenty-five or twenty-six of them, if I'm right. We're certified on that aircraft, and that's a manageable size fleet for our small company." I waited to see his reaction.

"OK, OK," he finally chuckled. "I'm convinced." I believe Jim could see that I was quite serious and, more than that, enthusiastic, and I think he decided to let me run with this one to see where it would end up.

"Let me know if and when you need any help from me." Like the rest of the team, I knew I could count on Jim if and when the time came.

The next few days were occupied with preparing a presentation I could provide to any interested parties. It described our HUD in detail, as well as its many safety and operational benefits. The presentation was geared toward Alaska's operation, and mentioned their bad fog cities (I included the NOAA chart). I had the A-Team and Jim review it for corrections and improvements. They then listened to me give the presentation, asked me the kinds of oddball questions a customer might ask, and corrected my answers where needed. We even did this over beers. I soon felt fairly comfortable with my pitch.

The following week I took my first very nervous step!

I knew Alaska Airlines was headquartered near the Seattle Airport, called Sea-Tac, but I'd never actually been to their facilities. I called to get the address and directions to their Flight Operations Center—it was located just south of the airport. As I drove up from Oregon to their building, I rehearsed my pitch, often out loud, and tried to second guess any of their possible objections.

Too expensive, don't need it, don't want it, small company, no customers, blah, blah, blah. And I worked hard to formulate an adequate response to each objection. I also hoped to uncover the reasons behind the terrible HUD conversation I'd had with Boeing before joining Flight Dynamics.

The receptionist in the lobby of the Alaska building was very polite. "Who do you want to see?" she asked, after I introduced myself.

"The chief pilot, I suppose," I answered with my fingers crossed behind my back. I was winging this.

"Do you have an appointment?" she continued quizzically.

After admitting I didn't have one, she asked why I wanted to see him. I explained a little about the HUD, but when I got to the "avoiding fog disruptions" part of my spiel, she held up her hand, stood up, and disappeared into one of the conference rooms close by. Returning a few minutes later and smiling, she told me that if I could wait fifteen or twenty minutes, a Captain Bruce Caulkins, their chief pilot, would give me a little time.

Gee, that was easy!

The receptionist then led me to their coffee stand, got me a cup, and took me into Bruce's office, which was very nice indeed. He had an excellent view of the airport runways from a huge wall window behind his chair. Around the room were reminders of his interests, life, and career. He had World War II aircraft paintings on the walls, family pictures on his desk, a nice Boeing 727 model painted in Alaska's colors, and a number of trophies, golf maybe, but I wasn't sure.

"What can we do for you?" he asked, walking in with his own fresh coffee, a little out of breath. Bruce was a serious-looking, heavyset man, probably in his late 50s. I introduced myself and we shook hands and exchanged business cards.

"I'm with Flight Dynamics in Hillsboro, Oregon, and I wanted to tell you about our HUD."

"Yeah, I've heard a bit about it," was his response. Well, I thought, that's encouraging, and at least he hasn't told me to get the hell out of his office yet.

"And I've been reading some interesting aviation articles describing it. They are all quite complimentary," he continued. This was indeed great news, and encouraged me to keep going.

As I wandered through the little presentation that I had worked up before leaving Hillsboro, I kept a close eye on Bruce, watching for his reaction to the various elements I laid out before him. He asked a few questions as I took him through the briefing, and I felt good about being able to answer virtually all of them, thanks to my trial runs with the A-Team back at the office. Finally, we got to my "flight disruption avoidance during foggy conditions" chart and Bruce became quite intent.

"So, you really think this HUD of yours will keep us flying in the fog?"

he asked.

Was he kidding? Was this a test, a trick question?

"Absolutely!" I chimed. "It's FAA certified on the 727 to CATIIIa minimums." I was showing off again.

"Hmm, that is very interesting," he continued. "So what's the payback?"

The what?

I had thought about the many possible questions I might be asked by a potential customer but here, I drew a total blank.

Payback sounded like a mafia term.

"You know, how long does it take to pay for itself?" he explained, obviously dismayed by my confusion.

When really stumped, ask a question!

"Well, when you have a flight disruption because of the fog…" I was reaching, "what does it cost your airline?"

Bruce thought about this for a few minutes, giving me precious time to collect my eroded confidence.

"Not sure, but its gotta be over a thousand bucks," he thought out loud.

As we continued to chat about the flight disruption scenario and how our HUD could help his airline, I realized that there was a whole other element to selling this frozen turkey head that I had not considered. And it seemed to be quite critical.

Bruce was standing up.

"Look," he said, now rather hurriedly, "I've got another meeting, so I've got to run, but if you give me a call in a few days, I'll see if I can ask around here regarding the cost of our weather delays."

Yes! An invitation to re-engage. The Number 1 rule of sales and marketing: don't leave your potential customer without some follow-up action. And this was on him.

I promised to get back to him and almost floated out of his office.

Back in Hillsboro, it was difficult to wait the few days. I passed the time by explaining to Jim that we would need a "payback." Jim seemed to understand, but told me he had no clue how to go about it. We would need to do some serious work on this aspect of our frozen turkey head effort. When the few days was up, I decided to visit Bruce's office again rather than calling him on the telephone. Face-to-face was always better.

This time I made an appointment with his secretary.

And this time he saw me right away.

"Well," he almost laughed as we got settled into his office again. "After discussing this with some of our folks around here, it appears that we have no clue what it costs us when we get these weather disruptions. As crazy as that sounds, we just throw money at the problem 'til it goes away."

I liked the sound of this…a lot.

"By the way," he continued, "where can I see this HUD of yours?"

This was the question I really wanted to hear!

"We have it installed in a Boeing simulator just down the road in Renton." I was overjoyed and had a hard time containing my excitement.

"Can I make an appointment for you to fly it, Bruce?" I asked, a little too expectantly.

"Sure." After checking his calendar, he got up to go. "Call my secretary and schedule a session for me soon. I'm definitely interested in your HUD."

After providing him with a little hand-drawn map to the M-Cab simulator, I thanked him profusely as I left. I decided I couldn't wait to tell Jim, so I asked around, found a payphone in the building's cafeteria, and gave him the news. He was truly elated, and assured me that I was already farther than he had gotten with this potential customer.

I called Bruce's secretary and scheduled a simulator session for a couple days later. As the big day approached, Jim and I strategized about getting to the next step at Alaska, and how in the world we could possibly develop a "payback." This issue was still there, and although we both realized the importance of it, neither one of us had any idea how to solve it. But still, Jim seemed really pleased with our progress.

At last the big day arrived. We would be flying to Seattle. Early that morning, an excited Mike Gleason and I boarded an Alaska Airlines 727 at the Portland airport, and as the jet engine whine grew louder for takeoff, I felt as if our company's luck had finally turned. Mike and I had become good friends over the past few weeks, and after hearing about my Alaska discussions and progress to date, he shared my enthusiasm about our upcoming session with Bruce.

It was a short flight. About fifteen minutes in, the pilot came over the speaker system and announced that there was a problem. There was fog in Seattle, and we would probably face a delay.

Geez, here was irony!

But for the first time, I felt really involved in this weather disruption. After we'd been holding over Seattle for nearly forty-five minutes, I became worried that we might miss our scheduled rendezvous with Bruce in the Boeing lobby. I told Mike about my concern.

He had an idea.

He went up to the cockpit and asked to speak to the pilot, who, surprisingly, obliged. Mike explained the situation and our concern about missing our appointment with Bruce. So the pilot radioed Alaska's Flight Control folks to get the message to Bruce that we would be a little late—because we were now caught in foggy weather on one of their aircraft. Maybe this would work to our benefit in the end.

After holding above Seattle for well over an hour, our flight finally got clearance to land. We disembarked and raced to the airport car rental counter, where a panting Mike instructed me to buy a couple of sandwiches from a nearby vending cart, since it was going to be one of those days.

We got into the car, threw the sandwiches onto the back seat, and quickly made our way to the Boeing lobby. No sign of Bruce.

Damn!

We waited and waited. Still no Bruce.

Finally I called his office. His secretary informed me that Bruce would not be coming today, since there were airplanes delayed and diverted all over their route system. He was a very busy man trying to cope with the mess the fog had created. I explained the situation to Mike.

"Darn it. Well, at least we can have a good lunch." Mike had an uncanny knack of always finding the silver lining.

We headed over to the Black Angus restaurant, not far from the simulator building. After a decent steak sandwich and a glass or two of beer, we were ready to head home. When Mike opened the car door, he saw our "airport sandwiches."

"Golly, what do we do with them?" I asked.

"Watch!" was his reply.

He placed the triangular sandwich packages one in front of each rear tire of our rental car. We then got into the car and he punched the gas. We stopped to look at the damage, and a laughing Mike observed that the sandwiches had become part of the Black Angus parking lot.

He looked up at me, grinned his familiar grin, and exclaimed, "That's the Bruce Caulkins Memorial Sandwich."

An Alaska Airlines B727 taking off in "sporty flying conditions"

Chapter 4

A Guru Emerges

Early the following week, I decided that a return visit to Bruce's office was in order.

My primary objective this time was to reschedule our botched first attempt at getting him into the simulator. I decided on a drop-in. By now, I'd gotten to know Bruce's friendly secretary on a first name basis—she was Chris. I found her and asked for her boss. Bruce came out to see me right away and apologized profusely for the previous week's mess.

But the ugly fog episode had gotten him thinking about "this HUD thing" again, and he mentioned that he had asked his Assistant Chief Pilot, Captain Tom Johnson, to "honcho" this project for them.

This was great news! At Bruce's suggestion, Chris led me down the hall and into Tom's office.

Tom was a little older than Bruce, slim and extremely pleasant looking. He stood up, smiling, as we entered, and Chris introduced me. We shook hands and exchanged business cards. He invited me to sit down and we got into some small talk. Tom had also heard about the little company from Hillsboro that was developing a HUD. As I went through my becoming-more-familiar HUD-101 routine, Tom became quite interested, even more so than Bruce had. He asked good questions, and some I couldn't even answer. But I wrote them down and promised to get the answers for him.

I felt that this was a really promising step in the right direction at Alaska Airlines.

Tom went into great detail on Bruce's, and indeed the whole airline's, frustration over the recent fog event, and the repercussions to the passengers and their flight schedules. I mentioned that my colleague and I had been caught in the mayhem too. We also discussed the downstream disruption scenario and he shared with me some of the major issues facing an airline that gets hit with these kinds of issues. He was also anxious to get into the simulator, and as I left his office, very pleased with myself, I immediately headed for "my phone" in the Flight Operations cafeteria to call Jim. As it turned out, Jim already had a simulator session scheduled for the next morning.

"I can rearrange my agenda to get Tom into the simulator tomorrow," he offered.

I hesitated slightly, concerned about horning in on Jim's session, but he was insistent.

"Nothing is more important than getting our first customer, Phil…nothing!" he said, and told me he would alert Mike to be ready. I returned to Tom's office and gave him the good news.

"Excellent," was his enthusiastic response. "I'll be there."

I left him a small map with directions to the Boeing simulator building down the road in Renton. I could tell that Tom and I would get along very well.

Since joining Flight Dynamics, I had become a serious fan of the weather segment on the daily TV news. That night, after dinner at home in Everett, I found myself glued to the weather report with more interest than ever. No sign of fog tomorrow. I did not want a repeat of Bruce's failed session. I was elated about the progress we were now making at Alaska Airlines and really felt on top of the world.

The next morning dawned clear and bright. I grabbed coffee and practically ran to my car. I had to keep a close eye on my speedometer all the way from Everett down to Renton. Arriving at the Boeing lobby, I found both Mike and Jim waiting for me. Jim had decided he wanted to be there—he also felt that we were getting somewhere with this potential customer, and I could tell he was excited about our upcoming session. After all, Tom had been "assigned" to investigate our HUD. Surely that was a good sign.

Tom arrived right on schedule. I introduced Mike and Jim, signed our guest in, and we headed down the now-familiar hallways to our M-Cab, making small talk as we went. We introduced Tom to our Boeing operator and climbed into the cab. There were four seats in the simulator so we all managed to squeeze in. While Mike set up his repeater display and prepped the simulator controls, Tom got into the pilot seat and began to adjust the seat position to get himself into the "eyebox," the best viewing position for the HUD.

All HUDs have an eyebox, due to the nature of their optics. I had learned this from Bob Wood during my A-Team educational meetings. It's an imaginary three-dimensional box in space where both of the pilot's eyes must be in order to see all the HUD symbols. I remembered all the adjustments I had to make with the pilot seat controls on my first try, and wondered who the hell had designed them. But Tom had 727 experience, and got it right in mere seconds.

Mike closed the simulator door and began to go through his HUD intro class for Tom.

This was the first time I'd been back in the simulator as an employee of Flight Dynamics. As Mike went through his briefing, I found that his words now had a whole new meaning for me. I actually understood what he was saying about the system, and the little green symbols, and I vividly recalled my confusion during that first session. I was thrilled to "get" the whole thing now! After no more than five minutes of instruction and answering a few questions from our guest, Mike released the simulator and Tom was flying. In the silence that ensued, I realized that I had practically stopped breathing in order to listen to every word that might escape from Tom's mouth.

But Tom flew in silence to a perfect landing...without looking down once.

"Wow...incredible!" Tom's anticipated words as the aircraft came to a stop on the runway centerline reminded me of how I felt after my first attempt. "Let's do that again," he said.

Tom flew CATIIIas, takeoffs, black holes, and circling approaches, and I watched in awe as Mike and Jim worked their sales magic on our guest.

I could tell it was working. Tom was truly amazed at his experience with our system. As we walked out from the Boeing building after our successful session, we invited our new friend to lunch, and over a good meal at the Red Lion hotel just down the road from Tom's office, we heard why he was so interested in our HUD.

In his former life, he had been a copilot for a charter air carrier called Standard Airways. During a pre-dawn return to Vancouver, BC, from a Honolulu turnaround, they had encountered one of a pilot's worst challenges—ground fog. The control tower reported very poor visibility, but the flight crew on the approaching aircraft could clearly see the runway markings, lights, and airport beacon. The dispatcher on the ground reported stars visible from his location on the field and thought the poor visibility report must be a pocket of fog at that location. The plane was given clearance to approach for a check. Tom flew the approach while the captain kept the runway in sight. Everything looked fine, so they continued.

All went well until the speeding Boeing 707 touched down on the runway and the crew immediately lost all forward visibility. Since the fog was only about a hundred feet deep, the runway lights and paint stripes had been visible from above, but once on the ground and in the fog, they couldn't see a thing. As their machine decelerated, they suddenly realized they were off the runway.

Tom paused. We could see that this was a very emotional memory for him. We waited patiently.

He continued the story. The aircraft had rumbled over the rough airport terrain, colliding with some stationary aircraft parked near the airport hangars, until finally the crew could see some lights. They were the lights of a terminal building looming directly ahead. Just before the still-decelerating 707 crashed into the building, it struck a concrete parking curb, which sheared off the nose gear, causing the airplane to slide on its nose into the building. A large horizontal structural member sliced well into the roof of the airplane's cockpit, passing over the tops of the crews' heads by mere inches. This experience had left Tom badly shaken, and very concerned about the insidious nature of fog.

No wonder he was so excited about our product.

The very next morning, Tom called me at home in Everett. I remembered that this was another good marketing indicator, when the potential customer calls you. I was just fixing my coffee. He asked me if I could come down to the Flight Ops building to meet someone. "Absolutely!" I replied, and within an hour, I was sitting in Tom's office. Bruce joined us.

Yesterday's simulator session had definitely done the trick. Tom was hooked. He absolutely bubbled as he told me about how, on his return to their facility, he had briefed Bruce and Pat Glenn—VP Flight Operations for Alaska Airlines—about his HUD experience in the Boeing simulator. Tom explained that Pat would be key to any progress for our program there. Bruce added that he was very disappointed that he had missed his scheduled simulator session a short time ago and vowed to take another crack at it.

Holy Cow! VP Flight Ops. We were moving up the food chain.

I assumed it was Pat Glenn that Tom wanted me to meet. He stood up, indicating that I should follow him. We exited his side of the building, went across the center hallway, and entered a set of offices on the other side. Tom marched me up to a cubicle where one of his "worker bees" was quite busily attending to some paperwork.

"This is Dean Schwab," he announced, as Dean stood up.

We shook hands. Tom explained that he had briefed Dean on our system, his experience with the simulator, and our progress so far, and had asked him to lead the HUD payback analysis effort for Alaska Airlines.

"You two have a lot to chat about," he yelled back over his shoulder as he walked out and left us.

Once again, I found myself immediately liking my new contact. Dean was about my age and appeared eager to learn more about our HUD. I provided him a little background on the company and our M-Cab session with Tom, and then took him through my little presentation. That first day, we spent hours talking about the HUD and how we could put together a payback and a good argument that Pat Glenn could take "across the street."

That was their term for Alaska Airlines' Headquarters, a building about a half mile away.

There would be some serious work to do to get there.

Tom Johnson behind the wheel of the Flight Dynamics HUD in the M-Cab

Dean Schwab in the copilot seat of an Alaska B-727

From the Winnipeg Free Press, Manitoba, February 8, 1968

INVESTIGATION LAUNCHED INTO B.C. PLANE CRASH

Vancouver (CP) -- The wreckage of a jetliner, buried to its wings in a building after cutting a mile-long arc of destruction early Wednesday, has been pulled free for an intensive investigation. Two persons were dead and 18 others injured when the Boeing B-707 leased by Canadian Pacific Airlines crashed on landing in heavy fog.

The 100-ton plane veered off the runway to the right then skidded and slewed toward the main passenger terminal and tower facilities of Vancouver International Airport.

"We were lucky on this one," said Bob Cole, a transport department spokesman. "It could have been a hell of a lot worse. We had some miracles out there."

Passengers and witnesses said the plane, owned by Standard Airways of Seattle, seemed to bounce on landing, veer, catch fire in a forward section and then began hitting things. It first charged through a covey of four small planes, leaving them tangled bits of smoldering wreckage. Then its right wing sliced through a frame transport department field office. A fuel tank ruptured and aviation fuel fed a resulting fire. Several cars parked nearby were also burned. The jetliner continued on its mile-long wild arc toward the terminal. It dropped an engine and part of its landing gear.

En route it clipped an Air Canada DC-8 waiting to be loaded. Then it picked up an Air Canada ramp vehicle and dragged it along as the big plane buried its nose in bricks of the Aviation Electric Pacific Ltd. building, 50 yards from the terminal.

"I couldn't see any airport lights at all, and I remarked to my wife that the pilot would be doing well to bring us in with such darkness," said passenger CONRAD LAMB, of The Pas, Manitoba. "Then, bang, we hit."

The wrecked plane was the only B-707 in the CPA fleet.

The cockpit crew were Standard Airways employees from Seattle and were identified as:

Pilot Captain AL BURKHALTER.

Co-pilot TOM JOHNSON.

Navigator CURTIS DAHL.

Engineer JOE NACENKA.

US federal aviation agency will join the transport department team today in the crash investigation. Transport Minister Hellyer told the Commons that his department had sent an investigation team to the crash scene. It was the first crash at the airport since Jan. 2, 1966, when a B.C. Airlines Grumman Goose twin-engine amphibian overshot the runway in heavy fog and crashed on tidal flats, killing three persons.

Chapter 5

The Twelve Days of Christmas

Christmas had always been my favorite time of year. This season, I was to receive a most unexpected gift...and some very bad news.

There were two high priority activities in the weeks that followed my successful initial efforts with Alaska Airlines. The first was my continued education on the hows and whys of the HUD. I'm sure my questions drove my Flight Dynamics colleagues crazy. We would talk about it incessantly—over lunches, happy hour beers, during coffee breaks, you name it. I couldn't seem to get enough. Still, my teachers were thrilled with my continued curiosity in the HUD, and my constantly improving knowledge of how it worked and its many safety and operational benefits.

My second priority had been triggered by the comment made by Bruce at that first meeting with him many weeks ago, when he'd asked about payback. Dean had assured me, after doing some internal research, that Alaska did not have much recorded data on fog episodes, the resulting flight schedule disruptions, or the costs of coping with them. This seemed to back up Bruce's earlier comment about not knowing how these events impacted the airline's finances.

I did some investigating myself. I found and read a number of industry articles about airline flight disruption scenarios. Most of these scenarios were related to mechanical issues, which was like a cog in the machine was broken but the machine continued to run. But a fog event shut the machine down altogether. I found some good, juicy information on the airline impact of flight disruptions, but there was still a helluva lot missing.

Christmas 1985 was to change everything!

It was a couple of weeks before Christmas. Back in Hillsboro, I walked the familiar Monday morning route from the lobby to my office. I ran into Vanessa, one of the women who worked in production. She looked like she'd just lost her best friend. Then I saw someone else in the same condition.

"What the hell is wrong with everybody around here? It's Christmas for God's sake," I said to Vanessa.

"Phil, haven't you heard?" she queried, looking very sad indeed. "It's Mike Gleason."

"What about Mike? What's wrong?" I was now quite anxious.

"He's leaving us...he's going back to Continental Airlines," she continued as she walked away with her head hanging.

No! That's impossible! Not Mike! Not now! Please, not now!

I dropped my briefcase onto my desk and almost ran to Mike's office. He was standing up and putting all his personal things into a cardboard box.

"Mike, what's this I hear?" I almost barked at him.

"Yep, its true Phil. Continental has resolved their labor dispute and have given all pilots just three days to get back on the job if they want to retain seniority. I've told John and Jim. Too bad, I was really enjoying working here, and with you on your Alaska project."

"But...but..." I couldn't find the words. Trying to do this HUD thing without Mike at the controls was inconceivable to me.

"But Mike, can't we change your mind?" I was desperate and really hoped our blossoming friendship might play a role.

"Nope." Mike continued to place his possessions into the large, overloaded cardboard box on his desk. "Need to get back to my family, too."

I slumped into the chair across from Mike's desk and with my elbows on my knees, I put my face in my hands.

"I can't do this without you Mike." I was pleading. "We're almost there on this Alaska thing."

"I know, I know, but I gotta go. I'm so sorry, Phil! Merry Christmas," was Mike's quiet answer as he approached from around his desk to shake my hand, apparently for the last time.

I could tell he was bound and determined. I was devastated.

As always in these difficult moments, I found myself on the way to Jim's office. He smiled as I entered, but I could see it wasn't his usual happy face.

"What the hell are we going to do now?" I asked, hoping that Jim had a backup plan.

"We've gotta move on, Phil. Tomorrow I am going to do some searching for a replacement."

A replacement? For Mike? It would be a tall order to find someone to fill his shoes.

Somehow I managed to get through the next few days in Hillsboro. I was upset about Mike's news, and concerned about how we would get to any of the next steps with Alaska. Tom had asked me to set up a simulator session for Pat Glenn, and another for Bruce Caulkins. Who would run those sessions?

Later that week I was glad to be back home in Everett for the holidays. As usual for this time of year, work-related things slowed down. Most of my Christmas shopping was done and I was looking forward to spending a brief, calm interlude with my family after the hectic pace at Flight Dynamics.

But the black cloud of Mike's departure hung in the background.

One evening, as I was relaxing with my family, watching the news after dinner and sipping on a glass of wine, the newscaster mentioned that Seattle's airport had been hard hit that day with exceptionally heavy fog, and there were many flight delays and cancellations for all the airlines. They even interviewed some of the stranded and very annoyed passengers in the waiting lines at the airport. It was what they said that sparked my interest.

The passengers were far more upset than usual because of the time of year, and I realized that most folks who were travelling at that time were going somewhere for Christmas: home, or to visit friends, family, and relatives for the holiday season.

By now, Tom and I had become good friends. Although it was about 7:30 pm, I decided to call his office.

Tom was still there. He answered right away and sounded extremely flustered. I had never heard him flustered in all the time we had spent working together over the recent weeks.

"We've got airplanes everywhere and none of them are where they're supposed to be," he almost yelled. "Seattle is totally shut down and they're predicting the same conditions again tomorrow," he continued, out of breath. "We've got a major problem here, Phil."

"But Tom, why is this one so bad? Don't we have lots of fog occurrence in Seattle?" I fully expected him to hit me with "tis the season."

However, he explained that the reason for the total chaos was that, although fog was fairly common in Seattle, it was usually gone by mid or late morning and the airline's operations had the rest of the day to catch up. But today, the fog had hung around all day. Alaska had gotten very few flights in, and all of those were very late. As a result, they had diverted many of their airplanes to Paine Field up in Everett. The rest were "doing donuts" above Seattle, had diverted to Portland or Moses Lake, or were scattered about their system, and nearly all were in the wrong place.

Was this fortuitous?

Very early the next morning, after grabbing my coffee to go, I found myself racing down the I-5 freeway. When I got to Lynnwood, about halfway to the airport, I entered the heaviest fog bank I had ever seen. Slowing down to about 20 mph, I could barely make out the paint stripes between highway lanes ahead of me. More than once I had to slam on my brakes as I came up on a slower car.

The image of the gray screen on my first CATIIIa approach in the simulator with Mike came back, and I wondered how pilots felt when they didn't have a HUD.

My sadness over Mike's departure also returned.

It took quite a while, but finally I arrived at Alaska's Flight Operations Center. From the parking lot I could barely make out the now-familiar building outline through the murk. As I entered the facility, I felt as though I had walked into the middle of a full-blown tornado. I had never seen anything quite like this. People were scurrying around in a manic rush. There were pilots everywhere, filling out paperwork, managers barking instructions, telephones ringing off the hook. I bumped into Bruce as he was racing to a meeting.

"Did you pray for this?" he asked accusingly. I thought I detected a hint of a smile as he whisked by me.

I was getting a first-hand glimpse of what goes on behind the scenes of a major weather disruption, and it was not pretty.

I navigated the chaotic conditions until I found a very weary-looking Tom at his desk with a huge pile of paperwork in front of him.

"Not now, not now," he exhaled loudly as he reached for the telephone. I backed out slowly and headed across the hall to find Dean. Like the rest of the crew, he too was preoccupied with yelling instructions into the phone. I waited for a few minutes until he hung up. He looked really tired and I figured he'd probably been at his desk all night.

"Unbelievable!" he exclaimed as he looked up, saw me, and slouched back in his chair, utterly exhausted. "We've got fish in an aircraft from Alaska thawing out in Everett, and we can't find any more refrigerated trucks to get 'em to Seattle. We've got more passengers in the wrong place than in the right. We don't even know where some of our aircraft are. This is the worst I've ever seen, Phil. We never got yesterday's airplanes sorted out and here we go again today. The forecast is for fog all day, and now they're saying we might even have it again tomorrow. Geez…is this a good argument for your HUD?"

His words struck a chord. He was absolutely right, but how could I get my arms around it?

Dean told me that he planned to record all their flight disruptions to help us with our HUD payback analysis. He had already listed, as much as possible, every flight and what he knew had happened to it since the start of the fog episode. He was also collecting receipts for all the extra fuel being consumed for the flight holds, diversions, and delays.

"But the real mess is at our Station Operations Center at the airport," he continued. "They're the ones that take the brunt of all the passenger heat."

"Really! Can I go over there?" I thought I might as well get the whole picture.

Dean looked at me like I was crazy.

"What the heck for?" he asked incredulously.

"Well…," I answered, "I need to see things first hand, and if that's where the worst impact is felt, then that's where I need to go."

Dean picked up his phone and called Dennis Kelly, who was in charge of Alaska's Sea-Tac Airport Station Operations.

Hearing only one side of a telephone conversation is a bit tricky, but I could tell Dennis was definitely in no mood for company. Dean explained Alaska's interest in the HUD and promised repeatedly that I would stay out of his hair. Finally, after some dickering, it sounded like we got the OK.

Dean drew me a little map of how to find the ops center in the airport terminal and threatened my life if I got in the way. I promised to be careful and headed out. The conditions outside had not improved. In fact, it looked far worse than when I had entered the building only a short time ago. As I got into my car in the parking lot, I could no longer make out even the faintest outline of the familiar Flight Operations building.

Driving half a mile down the road was now an incredible challenge. I could only see about one car length ahead of me. It was scary as I entered intersection after intersection, unable to see the red traffic lights. My heart raced as I narrowly missed colliding with other panic-stricken drivers. It took me nearly 30 minutes. But I knew the route well and finally parked in the airport garage and entered the terminal.

What a sight!

I had been involved in fog episodes before, even just a couple of weeks earlier as we had tried to get Bruce into the simulator, but I had never seen anything at all like this.

People were everywhere. Lots and lots of them. Sleeping on benches, on luggage, on the floor. Wrapped Christmas presents were strewn all over the place. Other people were engaged in annoyed exchanges with airline personnel assigned to "take care of things." It was total chaos.

Every available pay telephone had a long line. I tripped over a small pile of luggage as I made my way through the terminal. It was the Christmas presents that really hit me, and I immediately felt sorry for all these stranded families. Dodging the crowds of passengers, and following Dean's little map, I finally found the Station Operations Center. There was a phone on the wall outside the locked office door, so I called, asked for Dennis, and finally got inside.

Dennis looked even worse than Dean. Much worse! It must have been the aggravated passenger heat. He obviously had been there all night dealing with irate flyers, and annoyed company buyers who had not received expected goods.

He did not look at all excited to see me.

I introduced myself and heartily promised to stay out of his way.

Removing a small notebook from my briefcase, I began to write furiously. Between sorting through some of his paperwork, answering his telephone, and barking instructions to coworkers around him, Dennis told me about renting all the buses they could find to transport passengers from Everett, Portland, Moses Lake, and other diversion locations to Seattle and then back again for passengers wanting to get out. I wrote.

They needed more trucks to move cargo to and from Seattle. I wrote.

They needed more buses for passenger transfers, as well as vans to take their luggage. I wrote.

Three loads of seafood from Anchorage had already gone bad waiting for refrigerator trucks. I wrote.

They needed more taxis to get fresh pilots out to the remote locations so they could continue what was left of their devastated operation. I wrote.

There were multiple and various rental charges for support services at their diversion locations. I wrote

They were all out of hotel space. I wrote.

Pilots were running out of allowable flying time. I wrote.

Somehow, that day simply flew by. By seven o'clock that evening, I was as exhausted as Dean and Dennis.

And my little notebook was filling up, fast!

In between crises, I chatted with some of Dennis's coworkers to find out how much the airline was being charged for the buses, the trucks, the support services, the taxis, the hotels, and what was the value of the spoiled fish? It was real-time HUD economic analysis.

And it was simply incredible.

I vividly remembered Bruce's comment about "just throwing money at the problem until it was resolved." Here was a perfect example. He was dead right!

Much later that night, finally back at home, I grabbed my warmed-up dinner from the oven and was once again glued to the weather report. The prognosis for the next day was dreadful. The newscaster's TV camera panned the airport scene I had witnessed earlier in the day in person, and zoomed in on passengers sleeping on the floor, huge lines at the airline counters and telephones, piles of luggage and wrapped gifts everywhere.

I began to wonder how long this would last. How long could Alaska Airlines continue to hold on with less than ten percent of their flights arriving at their primary hub, and with virtually all of these arriving very late? Was this to be an endurance test for the airline?

Day three of the fog episode dawned much as the last two. I got up, tried to explain things to my very disappointed family, grabbed coffee, and bolted out the door for the drive to the airport. I laughed as I asked myself why anyone in their right mind would opt to go to the airport under these conditions if they were not trying to catch a flight.

Half way there I entered the dense fog bank, slowed way down, and tried not to get into collisions with slower drivers. I finally got to my new hangout and slumped into the chair across from Dean.

Dean had managed to get home very late the night before and catch a few Zs. He looked somewhat refreshed, although still quite frazzled. He was struggling to schedule pilots running out of air time, most of whom were in the wrong place. He was also trying to hold onto any buses and trucks they had contracted. Other airlines and freight carriers were in the same boat and also wanted any available alternate transport. It was the bus and truck company's dream come true.

And the forecast was for more of the same.

I showed Dean my little notebook and explained what I was doing over at Dennis's. I reassured him that I was not getting in the way, and that there was lots of juicy stuff I was collecting.

"Guess what?" Dean beamed at me after I'd finished.

"OK, what?" I asked, eager for a new lead.

"Apparently Eastern Airlines has been getting in every day since the fog started, and our management wanted to find out how. Well, it turns out that they service Seattle from Atlanta with an Airbus A300 that has CATIII autoland. Some of our passengers have figured this out and were actually flying out of Seattle on Eastern and then rerouting to their destination from Atlanta...even if they were going to LA."

Dean almost chuckled as he considered the circuitous route.

"Really?" I exclaimed. "Does that mean that CATIII would have solved our problem here?" My fingers were crossed. If I could just convince Alaska's management...

"Probably most of it," Dean replied, "but we'd have to see the RVR trace to know for sure."

The what?

After all my time with this thing, how could there be something else that I had absolutely no clue about? I knew that RVR was Runway Visual Range, a measure of the fog density along the runway. I had learned that from my A-Team training. But what was this about a trace?

"The RVR trace," Dean continued, "you know...from the tower."

Seeing my total confusion, Dean patiently explained to me that the airport control tower had a machine that recorded the RVR on the runways. Every airport had one.

Another light went on!

"How can I get there?" I again asked Dean for help.

He laughed heartily as he pulled a small book from his desk drawer. "I'll check for you."

Soon he was on the phone with Don Hughes in the NOAA office in the Seattle airport tower, and explained what we were up to. This time when he asked if I could visit, there was an enthusiastic reception.

Another map and I literally flew out the door.

The airport terminal was no better than the previous day. The only difference I noted was that there were now more TV station camera crews filming the chaotic conditions. I followed Dean's map, dodging groups of passengers and heaps of luggage, and arrived at an elevator that took me up to about the middle level of the tower.

Again a locked door, a phone on the wall, and Don arrived to show me in.

Again I was struck by the pleasant and cordial nature of so many folks in the aerospace industry, and wondered if sales people in other industries could say the same thing about their marketplace.

A very friendly Don led me down a passageway and into the offices, chatting about the fog devastation at the airport. But here, it was calm. No panic. Don went through a short explanation of the various pieces of equipment in the facility and what they did. Then he got around to the RVR machine.

It was quite simple really. A paper roll that looked for all the world like a roll of toilet paper fed into a machine with a scribe that inked a trace.

"Here's the problem," Don explained, pointing to the trace that now hung out near one side of the paper.

Don unraveled part of the roll. The trace for the past four days had dipped from two thirds of the way across the roll to very near the bottom.

"Right here is where the fog hit us." Don pointed to the sharp dip in the trace four days ago. "Here we are right now. And it's still with us"

"So what's the RVR right now?" I asked, gazing at the trace and the mechanism.

"'Bout 800…looks to me like." Don had a funny way of speaking.

"Wow, this is great!" I was ecstatic. And I was again fascinated and impressed by the technology we used to ensure the safety of air operations.

"How long does a roll last?" I continued.

"Oh, 'bout a month," Don said as he opened a closet and showed me shelves of rolls. On top, new ones, down below, used rolls with the dates stamped on them.

"What do you do with the used ones?" I was trying to contain my mounting excitement.

"After about four to six weeks, we can just chuck 'em," Don replied. "We have to keep 'em that long in case the FAA or the NTSB has an incident that they want to go back and check on."

"Well, can I have this one that's on the machine now, when you're done with it?" I begged, trying hard to contain my enthusiasm.

"Don't see why not. I'll check for you," he said, as he reached for the phone.

A short conversation followed and he hung up. As I explained why I wanted the roll, Don smiled and told me to call him around the end of January in the new year and he would get me the roll.

I felt like *I* was on a roll.

The next few days were blurry. The terrible fog continued with only slight reprieves, but the evening news was getting all too familiar. Newscasters were now recommending that folks drive even if they had a long way to go. The scenes of passengers sleeping on the floor and on piles of luggage were commonplace. The anchor folks were saying this was the worst fog that had ever hit the Pacific Northwest. Ever!

And what a time of year to hit.

In the days that followed, I continued my trips down to the Flight Operations building and the Ops Center. I had discreetly asked some of Dennis's folks to keep track of their various expenses for me. Each morning, many of them handed me a long list of charges and payments. It was essential that I capture as much of the ongoing damage as possible. Tom, Dennis, and Dean continued to struggle with airline, aircraft, pilot, cargo, and passenger issues, and I continued to write notes in my little books—I was now on number three. Hotels were booked solid all the way from north of Everett to well south of Tacoma. There were no more rental cars available anywhere. It was total chaos!

Meanwhile, Alaska's aircraft needed routine maintenance, but their main base was Seattle, which was now virtually shut down. I realized that all the careful planning that went into operations, routing, servicing—everything went out the window during such an extended period of disruption.

The final tally was twelve days. The now-famous Twelve Days of Christmas!

Seattle, Washington, sitting in a fog bank

Alaska Airlines B727-200 "holding" over Puget Sound

Driving in foggy conditions from Everett to Renton was a significant challenge!

Chapter 6

Across the Street

Personally, it hadn't been much of a Christmas season for me or my family.

But professionally, it was great!

Much of my Christmas holidays in 1985 had been spent crawling down the I-5 freeway to the airport (the dense fog turned a one-hour drive into three), chatting with Dean and Dennis, asking lots of questions, writing in my notebooks, and driving home long after dinnertime, often after my kids had already gone to bed. I had to explain to my frustrated wife the incredible significance and importance of what I was doing. I promised to make it up to them.

But now, early in the new year, I was back in my office in Hillsboro trying desperately to sort through all my notebook scribblings. There was a lot there if I could just disentangle it all. Much of it had been written in a mad rush. The coffee and food stains on many of the pages reminded me of the chaotic conditions I had experienced while I was there.

John Desmond and Jim Gooden were really excited about my RVR trace find. Along with my copious notes, it would give us a good picture of what had really happened to the airline over the disastrous Twelve Days of Christmas.

Jim had had no luck yet finding a replacement for Mike and there were pending simulator sessions for more of Alaska's flight operations folks: the Managers of Flight Training and Flight Standards, Fleet Managers John Powis and Kim Kaiser, and others. Our champion inside, Tom, had even managed to convince Pat Glenn, after the terrible fog episode, that he ought to take a look at the HUD in the simulator, and we had scheduled a session for him. I was worried that we were now facing a major project slowdown until we could conduct HUD simulator sessions for some of these key Alaska personnel who were critical for moving the project forward.

I was sitting at my desk in our office early one morning and suddenly heard a big ruckus outside our office door. I saw people running to the cafeteria. I jumped up and stopped one of the runners.

"What the hell is happening?" I asked.

"The Space Shuttle just blew up," the poor fellow responded.

"What? Impossible!" I retorted, as he tore himself loose and continued his run.

After dropping my notebook onto my desk, I joined the throng hustling to the cafeteria, where we had our only TV turned on. There on the screen was the shuttle's exhaust trail heading up into the wild blue yonder, but it did not look like the typical exhaust trails I had seen from other Space Shuttle launches.

As everyone watched in absolute horror, the newscaster explained that the shuttle, Challenger, had blown up shortly after launch. It sure did not look like there would be survivors. I suddenly remembered that this was the launch to send a school teacher into orbit. I had been a longtime fan of the space shuttle program and of NASA's efforts to get into space. I was devastated. This was terrible news. Vanessa, the coworker who had told me about Mike's departure, stood beside me watching. She leaned over and whispered, "We will all remember where we were when we heard about this one." I totally agreed.

This was a terrible setback for aerospace, and it deeply affected all of us.

One day, about a week later, I was at my desk again when a beaming Jim showed up. He had a new fellow in tow.

"Phil, this is Dick Hansen, our new pilot." Jim introduced us and we shook hands.

Dick was slightly older than me, a bit shorter, and nice looking. He was anxious to get an update from me on the status of the Alaska program. We sat down and I mentioned that he would be very busy indeed with all of our upcoming simulator sessions. I also explained what I had been doing in Seattle over the Christmas holidays.

He was fascinated.

Over the next couple of days, Dick became busily engaged in his HUD education with the A-Team, much as I had been earlier. He had been the Pacific Northwest demonstration pilot for Piper Aircraft, and he had another big advantage over me. Before Piper, he had been a US Air Force pilot and had even seen some action in Vietnam flying an F-4 fighter…which had a HUD. So, he was already very familiar with military HUDs—how they operate and what they can do. It did not take him long to get up to speed!

Dick and I quickly became good friends. I found myself chatting with him about our system over coffee breaks and after-work beers. During lunch one day, Dick told me that the HUDs he was used to on fighter aircraft were "dumb." He explained that they had no "smarts" per se, they just displayed critical flight and weapons status information to the pilot.

But our HUD was smart. It processed raw data and made changes to the pilot's guidance based on what was happening to the aircraft. So he proposed we change the name to "Head-Up Guidance System," or HGS for short. This made total sense to me. After explaining to senior management the need to distinguish our product, the new HGS name was adopted by everyone.

While Dick got up to speed on our system by talking to the A-Team members, I was very busy. I had managed to retrieve the RVR roll from the SeaTac tower. We were now regularly referring to it as the "Toilet Roll." Don Hughes at the NOAA office said they didn't need it anymore and we could do whatever we wanted with it. On my visit to his office to retrieve the roll, he had given me an explanation of how to read it, and what some of the trace anomalies meant.

Back in Hillsboro, I found a huge piece of white cardboard in one of the back rooms. I cut the RVR roll into twelve one-day strips and pasted them in sequence down the cardboard. I labeled each day with the date.

Next, I obtained the Alaska Airlines flight schedule for December from Dean. I traced vertical lines across the strips which represented the scheduled arrivals and departures for each of the twelve days of fog. Above each line I included the Alaska flight number. Since I knew Alaska was a CATII airline—that is, capable of landing in 1200 RVR—it was fairly simple to determine whether or not a particular flight would have successfully completed its operation. Takeoff required only 600 RVR, so the key was to concentrate on the arrivals. After all, if an aircraft couldn't land, this meant there wasn't one there for takeoff on the scheduled subsequent flight segment. If the RVR trace was below 1200, the line was red. If above, it was green. When I was done, well over half the flights were red. Alaska had scheduled about three hundred flights into Seattle over the Christmas period and only about a hundred and thirty of them had actually gotten in. Virtually all of those were very late to extremely late, making a total mess of the outbound flights and the downstream passenger connections.

But the interesting thing was that another hundred and thirty-five flights fell between their CATII minimums and the HGS's CATIIIa landing limit of 700 RVR. In other words, Alaska could have easily doubled the number of flights that operated, and most of those might have been on time or close to it. That would then have supported continuing departures from Seattle without major downstream disruptions and missed passenger connections.

This was real meat!

When I had completed the task, I invited John, Jim, and Dick to take a look at it.

They were impressed, and I could tell from Jim's expression that he was genuinely pleased with this step. It would be fairly straightforward to take this fog event and, using our NOAA statistical data, extrapolate the poor visibility damages to a whole year. By using my notes as baseline costs for typical fog disruptions, we could come up with an annual estimate of Alaska's fog damage. We would also be able to determine the percentage of cancellations, diversions, and delays over the year. Relating that damage to the cost of the HGS and installation in the Alaska fleet would yield us the magic payback period that Bruce had asked me about.

Other members of the A-Team had heard about my “Twelve Day Board” and visited my office to check it out and hear the explanation. They all seemed to sense that this was a critical step with the Alaska campaign, and they became as excited as our management.

It was now mid-February, and Jim wanted to get our findings up to Alaska Airlines as soon as possible, while the Christmas nightmare was still fresh in the minds of the airline’s executives.

I called Tom. After explaining to him what we had done with the RVR trace, he also became quite excited.

“Who at Alaska do I present this to?” I asked.

“Let me have a look at it first,” he countered, “and then we’ll plan an executive briefing.”

An executive briefing! This sounded very positive.

Once again, I felt on top of the world. John and Jim were excited about where this was going, and Jim mentioned that John had also briefed Geiger, our president, who also was becoming excited about the possibility of finally signing up a real airline customer.

The following week, while Dick was busy flying key members of the Alaska Flight Ops group in the simulator—including Pat Glenn, their VP—I drove up to see Tom with my Twelve Day Board. I drew a lot of attention, struggling with my prize under my arm as I entered the Flight Ops building. People and pilots passed by with crooked necks as they tried to read what was written there.

I finally got it into Tom’s office.

He called Dean over and the two of them looked intently at my board and listened as I explained how we did it and what all the costs were for the various disruptions. I had used Dean’s list of which flights had been diverted, delayed, and cancelled—my notes clearly showed there were significantly different costs for each. For example, in a cancellation, the aircraft’s fuel is actually a savings, not a cost, since the flight did not operate. In a diversion, there is more fuel consumed as the aircraft typically holds over, or near, the intended destination, and then goes to a new diversion destination as it runs low on fuel. Then there were the delays. I referred to them as either air delays requiring additional fuel, or as ground delays, usually with the aircraft engines at idle or even switched off, conserving fuel. So, I had to be very careful to include all the variables correctly in my analysis.

“So, what’s the final tally?” asked Dean.

“Out of three hundred scheduled arrival flights, you got a hundred and thirty in, and virtually all of those were very, very late. The HGS would have gotten you another hundred and thirty-five and those, along with the hundred and thirty you got in, likely would have been on time or close to it, supporting your continued downstream operations." I was really beaming now.

"So what was your calculation about the total cost to our airline?" Dean continued.

"My estimate, after doing all the homework, looks to be about two-point-six million dollars over the twelve days," I said. I watched for signs of disbelief in my little audience, but there were none.

"Wow!" Tom exclaimed as he slouched back into his chair. "That's unbelievable. Phil, you've got something here...a real business case. I'll set us up for a presentation across the street as soon as possible."

Across the street...YES! I called our office. Jim was ecstatic. This is what we had waited for. We'd only get one chance, so it had to be perfect.

To my surprise, Tom called a day or two later to let me know he had gotten a briefing window across the street rather quickly, and we were "on" for Monday afternoon of the following week. All the executive heavyweights would be there, including their president, Bruce Kennedy.

Needless to say, I sweated the next few days. I went over and over my briefing trying to think of audience questions that might come at me, and formulated my answers. Over the weekend, I found myself giving the presentation in my daily shower, and then answering all kinds of questions from fictitious listeners and arguing my point with them. My family thought I'd lost it. Sometimes even I wondered.

Finally, the big day came

John and Jim had decided they ought to be there to support me. Tom and Dean from Alaska had also decided to be there. We entered the main conference room at Alaska's headquarters and set up my Twelve Day Board at the head of a very long, very intimidating mahogany table. But we faced the board backwards. A secretary entered and asked if we wanted any coffee.

How about a beer? I almost asked, but stifled my request.

John and Jim were also nervous and this didn't help me at all.

Alaska's VP of Marketing, John Kelly, was the first to enter. Smiling and friendly, he immediately made me feel relaxed and comfortable as he laughed about this HGS changing the way we fly airplanes and making life easier for the airline's management. He was especially concerned about the effects of the now-famous Twelve Days of Christmas on the business passengers of the airline. From his vantage point, it had been even more of a nightmare.

Gradually, eight to ten other key members of the executive management team, including Bruce Caulkins and Pat Glenn, filtered into the room and took up positions around the table. Finally, Bruce Kennedy, their president, joined the group. I liked Bruce immediately. He was relaxed, yet commanded authority in a subtle way.

He announced that they were ready.

I began by introducing myself and John and Jim. As I went through my standard, but tailored, briefing, I watched carefully for any raised eyebrows or other signs of progress or potential problems. No questions were asked. Finally, I got to the Twelve Days of Christmas and the information the Alaska folks had provided me. After describing in detail all the efforts by Dean, Tom, Dennis, and Don Hughes, I reached for my "grand finale board," picked it up, and turned it around for all to see.

There was utter silence in the room as I described that here, in front of them, was a pictorial record of the fog event's damages to Alaska Airlines and its passengers.

Each red line represented an individual failed flight with its inherent direct and associated costs. When I showed the team that more than double the flights would have been completed and on time with the HGS, they were truly amazed. We had even been able to calculate the so-called ripple effect, that is, the downstream disruptions as a result of the fog issues at their primary hub in Seattle.

Finally, I put my pointer down in a gesture of closure.

Bruce Kennedy began the discussion.

"I recognized that we had a very costly event for our airline over this last Christmas season. So, after all this impressive analysis, Mr. Moylan, what was your conclusion regarding our total costs of coping with it?" he asked.

I looked around and Jim gave me a "go on, you're cooking" kind of look.

"Well…" I almost stammered, "by our joint efforts to assess the full impact to Alaska Airlines, we estimated the total to be in the vicinity of two-point-six million dollars for the twelve days."

I watched very carefully for any reactions around the table. No one said anything. The silence was deafening.

It seemed like a pretty high number to me, but after all the notes I had taken, and all of my analysis, I was confident we were very close.

"As a matter of fact…," Mr. Kennedy finally continued, "I asked Ray Vingo, our CFO, to investigate this event and come up with our own estimate of the total cost."

Uh-oh! I was taken aback. A nervous look flashed across Jim's face. Would Ray's number and mine be close? Would we be so far off that they might laugh at mine? I waited, sweating, for what seemed like an eternity, standing in front of the silent group of executives. Mr. Vingo stood up and slowly rifled through some papers in an open folder on the table in front of him as he fidgeted through his pockets for his reading glasses. With the glasses finally in place, he managed to find the right document.

"By our internal efforts here Bruce, we came up with…two-point-eight million dollars," he almost mumbled.

Two-point-eight! That was close! We were conservative by just enough, I thought.

I was ecstatic, and Jim and John were grinning ear-to-ear. Surely, this was a convincing argument.

There were a few more questions from around the table, but the point had been made. The money spent by Alaska Airlines on this event alone would have paid for our system on most of their fleet. As things around the boardroom table slowly wound down, some of the airline's executives milled around my Twelve Day Board, looking closely and asking me all kinds of questions.

"My wife was on flight 288 on Wednesday," said one. "Which one was that?"

I pointed to one of the red lines on the chart.

"Yep," he continued, "she ended up in Portland for twelve hours with our young daughter. Had to take a bus. Got home at 2 am. What a bloody mess."

I nodded, and smiled contentedly.

As we drove away from the headquarters building, I felt a tremendous sense of relief. Jim and John were visibly delighted with the way things had gone, and they were especially satisfied that our final number was so close to Alaska's.

I had done everything in my power to make this case. Hopefully it was enough!

Back at our office in Oregon, Jim and I updated Dick and the A-Team on the events "across the street" in Seattle. John briefed Geiger. Everyone was impressed and very excited.

The days following the briefing were tedious for me. My job was done. Now I had to wait, something that did not come naturally to me. How long would it take? What were they thinking? Were things going our way? Had we made the case? Did they need anything more? Was there a challenge to my analysis? Why had no one called?

Finally, I decided to drive up to Seattle again to visit my buddies at the airline. Tom and Dean had come up with a good explanation for the difference between our damage estimates and Ray Vingo's. While I had assumed their entire fleet of B727s were operating to their CATII limits, Dean explained that there were three aircraft awaiting some maintenance, which restricted them to CATI, or 2400 RVR, for the twelve days. So some of the aircraft I had assumed were continuing to operate properly were actually disrupted.

Meanwhile, another issue had been brewing in the background. I had known about it, but I'd felt it was secondary.

It was about to become primary.

Our HGS needed an Inertial Reference System, a so-called IRS. The IRS provided inertial data to the HGS to allow correct positioning of the flat-footed duck or, more correctly, the flight path vector—the actual projection of the aircraft center of gravity path through the air.

These were not cheap.

Honeywell and Litton both made systems that would work, but I thought it should be Alaska's decision.

Also, our engineers mentioned that the Air Data Computers (ADC) on the B727 were analog, but our system needed digital inputs. That meant we also needed two Digital ADCs, or DADCs.

These additional costs had me worried.

"Across the Street" was the Alaska Airlines headquarters

An Alaska Airlines B727-200 at the gate at Sea-Tac airport

OPPOSITE: Dean Schwab, left, and Capt. John Powis, two of the prime movers of the HGS program, review certification data. ABOVE: Flying the HGS in a Boeing simulator makes chief pilot Capt. Tom Johnson an enthusiastic HGS convert. LEFT: Even on a good day, landing at Sitka—as at many southeast Alaskan airports—demands precise energy management. Capt. Keith Kennedy flies the approach prior to HGS training.

Source: Air Line Pilot magazine, December, 1989

Chapter 7

More Equipment and a Big Oops

The full enjoyment of my win "across the street" was darkened a little by the realization that Alaska's B727s were older aircraft. I had not realized that they didn't have all the necessary modern avionics equipment to allow the HGS to do its magic. Each aircraft would need an IRS and two DADCs.

"So how much will all those cost?" I asked Doug at one of our regular meetings in Hillsboro.

"Probably will double the HGS cost per aircraft," he explained.

Oh no! I wondered if Tom and Dean at Alaska knew this was coming. I thought I vaguely recalled that Jim had explained it to Tom during our M-Cab simulator session, or maybe during the Red Lion lunch afterwards, but I couldn't remember for sure.

Once again, I did not want to wait until the famous Christmas season mess was a forgotten memory, so within a couple of days I was back at Dean's desk in the Alaska facility near Sea-Tac airport.

"I think we will need some other equipment to support the HGS program here," were the first words out of my mouth as I walked up to his cubicle.

But Dean, as pleasant as always, first congratulated me on our successful event across the street. For him, and for Tom and the Flight Ops team there, this was a crucial step in moving the entire HGS project forward. He had received very positive feedback from the airline's executives after our presentation of the twelve-day fog bout analysis.

He then explained that in follow-up meetings, both with Flight Ops executives and with their engineering department and across the street, Tom and Dean had explained that there would be some added equipment costs. In fact, Tom had already contacted both Honeywell and Litton to start the dialog on prices, options, and potential schedule for delivery of their latest technology IRS.

"Wow, that's great!" I said. I was very relieved, as I had been worried that we might have to develop another payback scenario to show Alaska's management, to justify the additional equipment.

"What about DADCs?" I was again proud of my increasing knowledge of related aircraft avionics.

"Not so good," was Dean's reply. "I've looked at some sources but can't find what we need."

Well I was sure as hell not going to face defeat at this stage of the game. We had come so far, and over a lot of really new and tough turkey head territory. I could now detect a glimmer of light at the end of the long tunnel.

I decided to turn to some of my contacts at Boeing. After chatting with a couple of the engineers and one procurement agent, I came up with a company name: Penny & Giles, based in Christchurch, England. The next day I arrived early at my office to allow for the time difference to contact the UK supplier. They were very cordial, and after I explained what we needed, they provided me the name of their US rep, Chris Thompson, who lived in Southern California.

I contacted Chris and explained our dilemma, and we decided to meet up the next week while he and I were both back in the Seattle area. Another lunch at the Red Lion Hotel. We quickly became friends and, yet again, I was reminded of what a great industry this is to work in. Chris was very personable and also very helpful. While he knew a little about our HGS, he did not know we had gotten as far as we had with Alaska Airlines. He laughed heartily as I explained what I had been doing over the twelve day Christmas fog episode. Then he explained in great detail why the DADC was needed, and why his was the right choice for Alaska. He provided me with some flyers about his company's DADC, and I provided him with the names of my contacts at Alaska.

After I gave them Chris's DADC brochure, the A-Team decided to write a spec for the equipment and invite two other companies to bid. After assessing all inputs, the team agreed that Penny & Giles was the optimal solution from our point of view. I was pleased about this outcome and called Chris.

A short time later, Chris told me that he had visited our buddies at the airline and they were all in agreement that his equipment would work for their HGS project. This was great news.

Later that week, back at home in Everett, I decided to drive down to Sea-Tac to discuss the program status with Tom and Dean. They were both thrilled about my P&G find and confirmed that they had met with Chris, worked with their own engineers to ensure that this equipment would all work, and included the costs in the proposed project plan.

Things were looking up again!

The next month or so seemed to fly by, and to our huge delight, Alaska decided the program was a go. Our CFO, Al Caliendo, went to Seattle a number of times to negotiate the terms and conditions of the contract with Alaska's executives. One day he arrived back at the plant and announced that the HGS contract with Alaska Airlines was signed.

For twenty-seven B727 aircraft, plus spares!

The A-Team was ecstatic, and I knew Geiger and our owners would be as well.

We had done it…our first real airline HGS customer!

There were many mini celebrations at the plant, and I ended up at a few interesting lunches and happy hours where we told funny stories about the toilet rolls and the twelve-day cardboard. I could tell that I was now a full-fledged and accepted member of our elite A-Team, as was Dick Hansen, the pilot who had replaced Mike. As for the A-Team, after many years of system development, prototype testing, and FAA HGS certification efforts, they were incredibly happy about this first big sale.

But a dark cloud appeared on my horizon.

I had been renting a small apartment in Beaverton, Oregon, not far from Hillsboro, to facilitate my work there. After the Alaska success, John and Jim had been pushing me to move to the Portland area so I could spend more time in the office. Boeing was still big in the plan, but they felt that this was going to be a long road and really wanted my presence at our plant and at our weekly marketing meetings on a more regular basis.

After discussing it at length with my wife, we listed our Everett house and began to shop for a new place in Oregon, close to the Flight Dynamics plant. We soon got an offer on our house, but it was quite a bit less than we were asking, and less than what we had paid for it a few years earlier. I explained this to our company management and suggested they cover the deficiency from my anticipated commission on the Alaska sale. My wife and I were not ready to "eat it."

Their answer was a resounding "no." My commission could not be paid until we received a large down payment from the airline in a couple of months. I tried unsuccessfully to convince them.

Quite upset, I explained this development to my very annoyed wife, and she strongly suggested I leave the company and come back to a job near Seattle. I tried yet again to convince the company's senior management to support my housing request, and while Jim and Dick and the A-Team pushed very hard to get support for my cause, it was not to be. John Geiger and Al Caliendo, the CFO, felt it would set an unacceptable precedent.

Two months previously, I had been approached by an aviation friend about a job opportunity with Bendix Avionics at the company's Boeing office in Tukwila, not far from Seattle. I called to see if he was still looking. He was. I decided to accept his offer, even though I knew I would forfeit the commission on the sale of the HGS to Alaska Airlines. I submitted my resignation, packed my things amid a lot of very long faces at the plant, loaded them into my car, and drove off. I also let go of my apartment in Beaverton, which I had been renting on a month-to-month basis.

It did not take me long to get accustomed to the Bendix product line. After all, almost all avionics were less complicated than the HGS. I settled into a nice office in Tukwila, and began to re-connect with a lot of the Boeing folks I had met during my Simmonds days.

About three months after joining Bendix, I got an interesting call from my frozen turkey head mentor, Jim Gooden. He mentioned that he and John Desmond were going to be in the Seattle area the following week, and wanted to invite my wife and me for dinner. Why not, I thought. Over a very enjoyable meal with my two former colleagues at the nice Anthony's Home Port seafood restaurant in Kirkland, not far from Everett, they updated me about progress on the Alaska front. They also asked how I was enjoying my new job and co-workers at Bendix. I explained that it was OK, but not as much fun as the HGS. And there was no A-Team!

After we finished eating, John leaned over the table and handed me an envelope. Inside, I found a check for my commission…the whole amount. I was amazed, since I had not expected to receive any. My wife was beaming too. John and Jim explained that Flight Dynamics had made a mistake in letting me leave, and they wanted to know what it would take to get me to come back.

I told them that the company would need to pay for flights down every week—I was tired of driving, and flying would allow me to spend more time in the office. Deal! Also my hotels, since I was fed up with paying the mortgage for my house in Everett, as well as rent and utilities for an apartment in Oregon. Deal! And rental cars. Deal! And a raise. Deal!

The days following that dinner were a bit tough. I sincerely apologized to my buddy who had offered me the Bendix job, for leaving so soon after starting. But he could tell I was more excited about the HGS than I was about their product line—I had talked about the HGS, a lot.

A scant two weeks later, I found myself once again in Oregon, but now at the new company facility in Tualatin, about a 15-minute drive from Hillsboro. They had moved while I was with Bendix. What a change! It was a big step up for the company. I was pleased that it was a lot more professional-looking, in case we had potential customers visit. I also had a real office, and while not quite as nice as the Bendix office, it was a big jump up from the broom closet I'd had in Hillsboro.

I had to hand it to John Desmond. While he and I did not always see eye-to-eye on everything, I understood that he had a lot on his mind and in his hands. He had held it all together, through engineering development, FAA certification, project logistics, and even financing, until we had managed to get our first HGS sale. He knew I loved my frozen turkey head job.

I was now able to attend regular sales and marketing meetings every Monday. At one of the first meetings, John asked what my next target would be. Having again anticipated this question, I had given it some serious thought. I told John, Jim, and the team, assembled and waiting with anticipation for my answer, that it was Western Airlines, based in Los Angeles.

"Why Western?" was John's question.

"Well," I responded, "they have a fleet of forty B727s, they do have some fog issues, and Mike Gleason has a friend who works for them and could help push the program, as well as advise me on who to contact there."

My answer seemed to satisfy the group. Later, Jim told me he thought it was a good choice, especially in terms of the size of their target fleet.

I was off to the turkey head races again!

Around this same time, Jim asked me to think about an advertising campaign. He thought it was time to let the world know about our HGS, its many benefits, and our Alaska Airlines sale. After thinking about this for a while, I asked John if we could hold a competition among the employees—we needed a good slogan. He agreed, and we decided to offer the winner a dinner for two at one of the nice restaurants in the area.

What a hoot! We got more than I bargained for. Some of the slogans were actually very good.

"This HUD's for You," a play on the wording of a TV beer commercial at the time.

"A HUD of our Time," which I especially liked.

But the winner, by group vote, was "The Future is Looking Up."

I put together our first-ever advertisement and showed it to John, Jim, and Dick. After suggesting a few minor changes and additions, they approved it!

After I'd submitted our new ad to a popular aviation magazine, Professional Pilot, the representative called to ask me for the name of someone at Alaska Airlines they could interview about the project and why they had selected our HGS. I gave him Tom Johnson's contact information, hoping that was OK with Tom.

After interviewing Tom, they asked him to actually write the article for them—something that was fairly uncommon at that time. Tom did a great job, resulting in a terrific article that even caught the attention of Boeing.

We ordered a whole lot of reprints and I ended up carrying quite a few in my briefcase as I ventured on to other targets.

I was, once again, a happy camper.

PROFESSIONAL PILOT

June 1988

Capt Tom Johnson, Chief Pilot of Alaska Airlines, was the company officer principally responsible for evaluating head-up displays for the carrier's weather-

A terrific HGS magazine article about Alaska Airlines, Tom Johnson, and the Twelve Days of Christmas

In Alaska Airlines cockpits, the transparent Flight Dynamics HGS-1000 HUD screen (or combiner) flips down when needed. The projector over the pilot's head casts a holographic image that keeps the pilot's eyes focused about 60 feet in front of the aircraft (in effect, at infinity) so that scanning for traffic and ground references does not require refocusing.

Capt. Kim Kaiser flying our HGS, now installed in a real B727 aircraft (source: Professional Pilot magazine, June 1988)

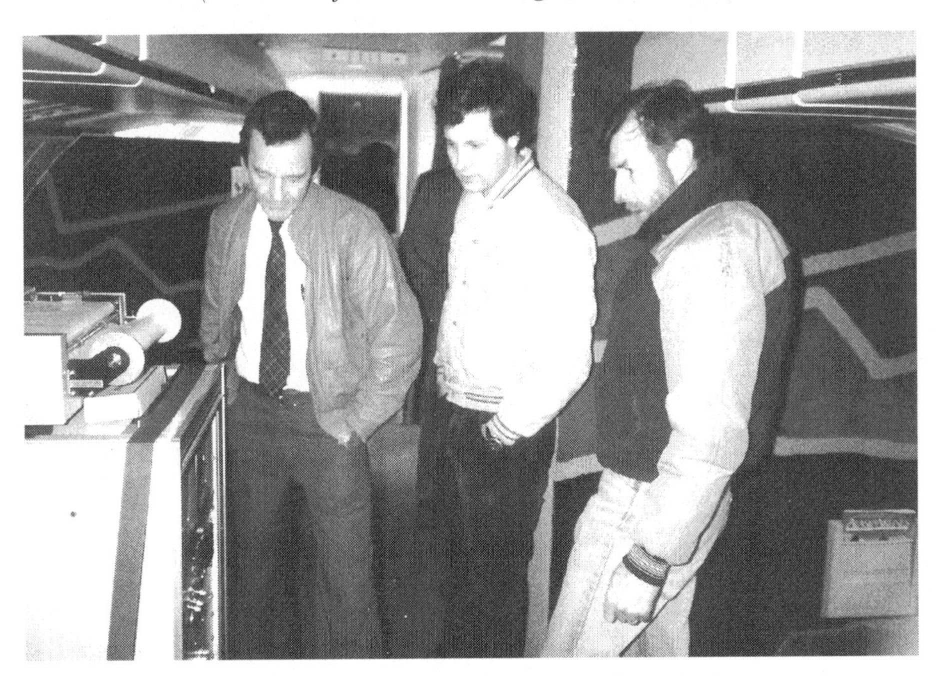

Berk Greene, Randy Foster, and Norman Jee checking HGS test equipment

The Future is Looking Up... at Alaska Airlines.

Fewer delays!
Fewer diversions!
Fewer cancellations!

An airport miracle? Not quite! But the holographic technology of the 90s will get Alaska Airlines' frequent flyers to their destinations....on-time....today....while others wait at the gate!

With the newly installed Head-Up Guidance System (HGS®) from Flight Dynamics, Inc. in their 727 fleet, this West Coast carrier expects a 70% reduction in weather related flight disruptions.

With windshear protection and head-up operations, the HGS offers Alaska's flight crews and passengers alike an added dimension in safety.

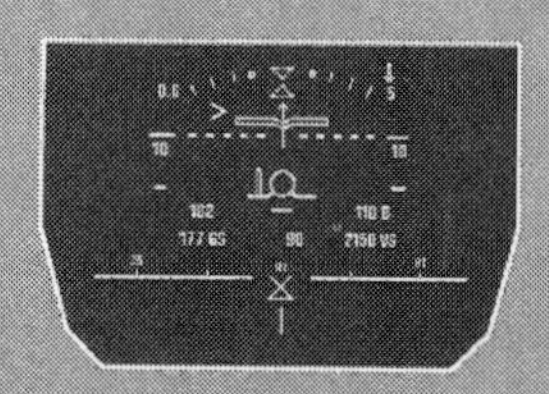

Let us show you how the HGS can improve *your* operations....and increase your profitability.

Contact us today for information, a presentation, or a simulator demonstration.

Flight Dynamics, Inc.
16600 S.W. 72nd Avenue
P.O. Box 230669
Portland, OR 97223
(503) 684-5384
FAX: (503) 684-0169
TWX: 910-460-2200

Our first HGS advertisement for aviation industry magazines

In my new office at Flight Dynamics in Tualatin, Oregon, I actually had a computer…my first!

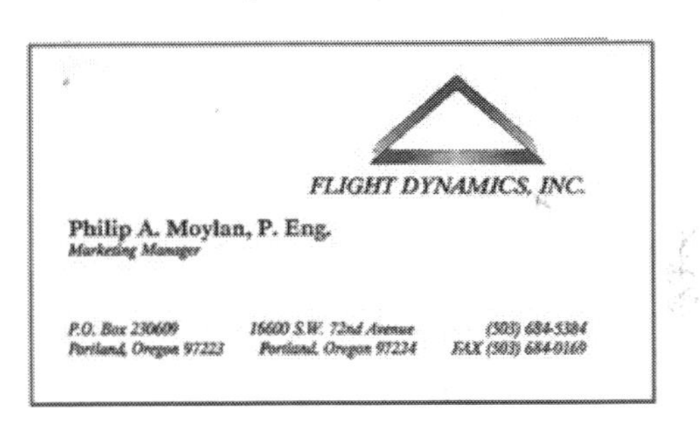

Chapter 8

New Target on the Horizon

Here is a story that is funny now, but wasn't so funny at the time.

During an experimental flight on one of the first Alaska B727 aircraft outfitted with HGS, just after takeoff in Anchorage, our small HGS cockpit annunciator control panel started to smoke. After a quick discussion about what to do, the pilots took out a screwdriver, removed the control unit from the cockpit instrument panel, opened the pilot-side window (they were still at quite a low altitude), and threw the smoking controller out the window and into the waters of the bay below.

When the A-Team heard about this we were all horrified. How could such a thing happen? Upon subsequent investigation, we realized that our HGS controller was located in the cockpit instrument panel immediately above the aircraft's Sperry weather radar unit, which was known for the superheat it emanated—in fact, you could fry eggs on the top of the unit in mere seconds. Our team came up with a solution—a protective "thermal blanket" was installed between the units, and we never had another issue like that again, thank goodness!

With my recent win at Alaska Airlines under my belt, I felt empowered to go after my next target with a vengeance. After all, I could now refer to an actual customer and provide real references. And I had experience answering just about every question imaginable about our system.

It was now late summer of 1986. Mike Gleason had been doing an HGS-101 job on his contact at Western Airlines in LA, and this contact had provided me a name: Captain Larry Hecker, who was VP of Flight Operations there. I called Larry and told him about the HGS and the decision by Alaska, which he had not heard about. He suggested I call him back in a day or two. Unknown to me, he then called Tom Johnson at Alaska to ask about their decision and rationale to equip their entire fleet with this HGS. So when I called Larry back later, I found a very receptive audience.

"How would you feel about providing our team at Western with a presentation?" he asked.

Of course I said I would love to.

"When were you thinking of coming down to see us?" asked Larry.

"How about Monday next week?" I did not want to waste any time.

I was rewarded with a resounding OK and a promise by Larry to gather the right team at the airline to hear my HGS pitch.

After relating my success to John and Jim, I asked if Jim wanted to accompany me. He suggested that, like with Alaska, I was better off on my own to plot out the initial strategy, meet key folks, and develop some relationships, and save Jim as an arrow in my quiver for downstream action.

Good idea. Sold!

The next week, I disembarked from a Western B727 at LAX. I figured it would be good to show up with a Western ticket jacket rather than Alaska. I found my way to the Western headquarters, not far from the airport. After checking in at their lobby, I was met by Larry's secretary, who took me upstairs to his impressive office and introduced me. Larry looked a bit like Kojak—a bald TV show detective at the time. He welcomed me enthusiastically, offered me a seat, and asked if I needed water or coffee. After some initial friendly chit chat and basic questions about the HGS, our company, and the Alaska campaign, he mentioned that Tom at Alaska had a lot of good things to say about me, the HGS, and Flight Dynamics.

In Western's main conference room, Larry had assembled an impressive team from Flight Ops and Engineering, and even had a procurement fellow there, which I took to be an especially good sign. There was Dewey Gerrard, Flight Ops Technical, Marty Farber, Head of the Pilot Union, and a couple of technical pilots and engineering types as well. After introductions, I took them through my now well-developed HGS presentation, which also included references to the Alaska deal and the Twelve Days of Christmas. I was very pleased that I was able to answer virtually all of their questions successfully, and with enough details to allay any fears they might have had. Even the additional IRS and DADCs did not hamper their enthusiasm. On the way out of the meeting, Larry introduced me to Robin Wilson, their president! I gave him a thirty-second HGS "elevator speech" and then he had to run to a meeting. Larry saw me back to the lobby and, patting me on the back, told me he was excited about the possibility of Western becoming a CATIII airline with our HGS.

Golly! It sure seemed like things were getting easier.

Over the next few months we had lots of work to do. Dick was busy flying some of the Western folks, including Dewey and Larry, in the M-cab simulator. I put together a payback based on the work we had done for Alaska, and included NOAA data for cities that Western served. Once again it looked really solid and, after John and Jim gave me the thumbs up, I submitted our analysis to the Western team.

Larry was extremely pleased, and told me he would make arrangements to brief Gerry Grinstein, their CEO, as well as their board and other members of the airline's executive management team, on the proposal and total cost estimate. The estimate was based on the per aircraft cost verified by Alaska, including the cost of doing the actual HGS installation. This "executive briefing" step had been vital for the launch at Alaska, so I was really buoyed up.

Dick and I had been working Western for about two months. Jim and John were quite pleased with our progress at the airline…until one day in September 1986.

That's when Larry called me.

"Phil, we have a major hiccup here at Western," were his first words.

"What is it?" I queried anxiously. "What's wrong?"

"I can't say what it is for a while yet, but I will call you very soon to fill you in on the details," he replied.

"Geez… Will the issue derail our HGS project?" was all I wanted to know.

"It might," he said. I went pale!

Jim and I went through a bunch of possible scenarios over the next few days, but finally, Larry called to let me know what was happening. Apparently, Delta Air Lines had made an offer to acquire Western, their board had accepted it, and as a result all capital acquisitions had been put on hold…indefinitely.

Including our HGS!

Later, after the acquisition, Larry accepted the position of Assistant FAA Administrator in Washington, DC, where he continued to tout the benefits of our HGS to the aviation industry.

Meanwhile, the A-Team was very disappointed about the Western news, but I knew there were other customers out there just waiting, so I tried my best to cheer them up. That's when the team started calling me "Captain Sunshine."

Our company now had a lot of folks flying around the country, including to Seattle for Alaska meetings and M-Cab sessions, and to Washington, DC, for meetings with the FAA. We had just completed a new HGS brochure, so I asked John if I could send a notice to all our employees asking that whenever they boarded a flight to anywhere, on any carrier, they pass an HGS brochure to the cockpit flight crew. He thought it was a good idea, and a great way to get the word out.

Around this time, John Geiger announced that he was retiring, and John Desmond would take over the president's position at Flight Dynamics. In reality, Desmond was already performing many of the presidential duties.

I was still reeling from the Western news when the telephone on my desk rang. It was very early 1987.

With fingers crossed that it was Larry with some good news from Western—which I desperately needed right then—I picked up the phone and heard the familiar and happy voice of my good friend Tom Johnson from Alaska.

"Phil, when are you going to be back up here in Seattle?" he asked.

"Absolutely anytime," I replied, since my current workload had just been significantly reduced. "Why? What's cooking, Tom…do you need anything from us?"

"No, but I have something that might be of interest to you and your company," he replied with a slight chuckle.

I perked up. "Tell me, what is it?"

"Sorry, Phil! You'll just have to wait till you get to Seattle to find out." He laughed his familiar laugh and hung up.

The next day we met at our now-favorite Red Lion Hotel restaurant. I managed to fly up, pick up my car, and drive over without feeling terrible about the turn of events at Western. I almost ran into the restaurant, I was so eager to see what was up. There was Tom with a tall, friendly fellow, whom he introduced as George Bagley, VP of Flight Operations at Horizon Air. Horizon was the regional feeder for Alaska Airlines. I was puzzled.

After the usual introductory chatter during lunch, George told me he wanted to investigate the HGS for his fleet of twenty-five Dash 8-100 aircraft. It took me a few minutes to get over my shock—this was a small aircraft to justify the cost of our system. But George explained his rationale to me. Horizon was responsible for feeding Alaska's B727 fleet for the longer distance flights that were carrying higher revenue passengers. Horizon was getting hit with all the same fog issues as Alaska Airlines.

He leaned across the table. "Phil, if we can't deliver our passengers to Alaska, their aircraft go out half empty."

He asked if it was possible to do a very high-level analysis of his route structure to see what might pop out. I promised to take a look.

After George left the restaurant, I told Tom I was extremely doubtful that our system on George's aircraft would provide a good payback argument. He offered to help me wherever he could. He also mentioned that John Kelly, who had been the first to arrive at our pivotal Alaska briefing "across the street," would soon be promoted to President of Horizon.

The stars were lining up, if I could only show payback…

I drove back to Everett, where my family and I still had our house. I could not imagine how in the world we could possibly justify the HGS for a thirty-seven-passenger Dash 8 aircraft when the B727 barely cleared the typical two-year required airline payback hurdle with a hundred-and-fifty-passenger load.

After arriving home, I called Jim at the office to update him. He was just as stunned as I had been. And he agreed that it would be an uphill battle to justify the HGS, since the Dash 8 also needed the IRS (though not the DADCs, as this aircraft came already equipped with those). Still, the addition of the IRS would greatly increase the per aircraft cost.

Back at my desk in Oregon a few days later, I began the task of looking into how we might justify the cost of our product on such a small regional turboprop aircraft. But after receiving Horizon's flight schedule from George, I suddenly became more encouraged. While the Alaska B727s would carry out six to seven flights per day, the Dash 8s were doing twelve or more 45- to 60-minute segments, especially with their new Portland to Seattle shuttle. I began to hope that the additional daily flights might make up for the smaller passenger load in our analysis.

In the meantime, Dick and I hosted George and one of his key technical flight operations folks, Perry Solmonson, in the M-Cab simulator. Just like Tom, they were both duly impressed and wanted even more to be able to justify the HGS acquisition. And then there was Tom Gerharter, Senior VP of Operations for Horizon, based at their Operations Center in Portland, who was also very impressed and became a staunch HGS fan as well as a good friend. His wife Penny and my wife had the same birth date, and we enjoyed an annual celebration together many times.

As we were busy looking into Horizon's schedule, George called me one day at my home in Everett to alert me to a sudden fog event in process. It mostly involved the airport in Portland, but also encompassed Seattle and some of the Eastern Oregon cities that Horizon served. Once again, I became involved in monitoring the airline's costs of coping. But this time, I had expert help from the Horizon folks, who really wanted our system, and I did not have to personally witness the ensuing bloody mess. After five or six days of foggy weather, we tallied up all their costs, and they were significant.

One thing I noted with Horizon, that had not occurred at Alaska Airlines, was their propensity to ferry empty aircraft. Since their flights were typically an hour or less, it was considered no big deal to just ferry an empty aircraft to the location where it was needed. But this added a cost to the overall event roster that was not there for the bigger Alaska aircraft, which had longer flight segments. When we were done, our analysis looked much better than either Jim or I had expected. I sent the results to a very pleased George Bagley. He felt that he would not have much trouble convincing the powers that be across the street (yes, Horizon had their own "across the street," right across from Alaska's), especially his new president.

A short time later, while we were still gathering all the bits and pieces of the Oregon fog event to extrapolate them to annual fog disruption costs, George called me again. After giving me the wonderful news that John Kelly had approved our Dash 8 HGS project, he mentioned that he was now in discussion with deHavilland Aircraft for some newer, upgraded Dash 8-200 aircraft, and he had told them he *must* have the HGS on these new aircraft from their factory—that was part of the deal. We had our second HGS customer!

This news was truly amazing, and would change aviation in a most unexpected way.

After I gave the great news to John and Jim, the A-Team—which now included Dick and me—went off for some congratulatory beers at our favorite drinking establishment. John was especially excited about the deHavilland news. Because it involved an aircraft manufacturer—an OEM—we were sure to learn a lot. (Maybe more than we bargained for!)

During our call, George had also provided some names and contact information at deHavilland. John Howarth was the Regional Sales Manager who looked after Horizon, and Steve Ridolfi, at the time, was Director of Marketing, based in Toronto, Canada.

After a lengthy chat with a friendly and very supportive John Howarth, I called Steve Ridolfi. He was less than enthusiastic. He explained that the addition of the HGS to Horizon's new aircraft order would not only need sign-off by his engineers and flight test folks, but they would also have to certify our system for the new aircraft. This would complicate the Horizon sales process…a lot.

But some behind-the-scenes activities were taking place that would grease the skids in unexpected ways.

Unknown to me at the time, our pilot, Dick Hansen, had been working with some folks he'd met from the German Luftwaffe (their Air Force). They wanted the HGS on their Canadair CL600 business jets, and had dragged the Montreal, Canada, aircraft manufacturer into the discussions. One of Canadair's chief technical pilots had actually come to Seattle to fly the HGS in the M-Cab with Dick. Not surprisingly, he loved it.

After returning to Canada, the pilot spread the word about the HGS, and recommended offering it on their brand new fifty-passenger Canadair Regional Jet, the CRJ, then in development. This would provide CATIIIa capability for any European customers of the new aircraft.

As it turned out, deHavilland and Canadair were now in the middle of an acquisition by a Skidoo manufacturer in Quebec called Bombardier, which planned to merge the two companies. The new organization was to become BRAD (Bombardier Regional Aircraft Division). So the Canadair pilot and deHavilland's Steve Ridolfi now worked for the same company.

The pilot contacted Steve, who mentioned that Horizon had requested the HGS system for a bunch of new Dash 8s. The pilot managed to convince Steve of the operational and safety benefits of the system, and his absolute and unconditional support for it on both the Dash 8 and the CRJ.

So Steve called our office one day soon afterward to ask me to come to Toronto to present the HGS to him and his Dash 8 technical team. Of course I readily agreed.

I had done some work with deHavilland while employed in Toronto for

Simmonds, so I knew where they were located. Steve's secretary escorted me up to their conference room, where I met Steve for the first time. He had invited a large and diverse crew to attend my HGS presentation. A big issue that came up at the meeting was the HGS certification on the new aircraft—deHavilland had no experience with CATIII regulatory approval.

Earlier, we had promised George Bagley at Horizon that we would include an FAA Supplemental Type Certificate (STC) for his aircraft, like we had done for the B727. So when the certification issue came up at the Toronto briefing, I asked the group, "What if we give deHavilland our STC at no charge?" The engineers replied that this would go a long way toward their endorsement of the system on the new Dash 8-200 aircraft for Horizon. It would also ease the sales process.

But we were faced with some significant HGS changes for the BRAD Dash 8-200 and the CRJ. These new aircraft were being developed with "glass cockpits," meaning the primary flight display, or PFD, was a CRT that could be configured to provide all necessary flight information on a single, large display for each pilot.

The Alaska B727 and the earlier Horizon Dash 8-100 aircraft were called "round dial" aircraft—the PFD was mechanical and additional flight information was provided on individual round dial instruments located on the panel close by. As a result, there were no specific requirements for the HGS symbology format on these older aircraft.

With the advent of the glass cockpit aircraft, BRAD was adamant that the HGS symbology had to emulate the head-down PFD symbology for continuity and safety. This was a big change for our system.

After a lot of symbology and other design changes, and extensive development efforts by the incredible A-Team, our company delivered the first HGS system to Horizon, who then installed it on one of their existing Dash 8-100 aircraft for certification flight testing. Dick Hansen was heavily involved, as was Perry Solmonson from Horizon, who also got Dean Schwab from Alaska to help.

The system was finally FAA certified, and we began deliveries of the HGS for Horizon's existing fleet.

We modified the STC to also cover the newer Dash 8-200s, and delivered the STC to deHavilland, who then delivered the new Dash 8 aircraft to Horizon—with the HGS installed.

This OEM installation was to become a significant and aviation-changing event. And we had our first OEM HGS customer!

At Flight Dynamics, we had focused on the "big iron"—aircraft like the B727—and we hoped, one day, to get the system on Boeing's 737s, 757s, and 767s. But the win at Horizon, and subsequently, at BRAD, had far more impact than any of us at Flight Dynamics had anticipated. We had believed that smaller turboprop and regional aircraft could not support the cost of the HGS and other required equipment. But these smaller aircraft had no autoland option at all for lower minima operations. Especially in Europe, where this capability was vital, the HGS win with the Dash 8 and CRJ had opened the door for many regional carriers to improve their operational capability. We were to gain many more wins on these two aircraft types, all thanks to Horizon's pioneering efforts with us. Everyone was happy!

One evening in the spring of the following year, a good friend invited my wife and me to a cocktail party in Bellevue, not far from my original Simmonds office. As I was sipping on a glass of wine and chatting with my friend and my wife, I happened to overhear two men conversing about our HGS. I eavesdropped. One of the gentlemen told the other that he was a 727 captain for Alaska Airlines. He pulled a keychain from his pocket and I had to keep from bursting out laughing.

I had earlier developed the keychain for Flight Dynamics. It was clear, soft plastic with our HGS combiner shape and all the green symbology on it. The Alaska pilot asked the other fellow, who turned out to be a Dassault F2000 business jet pilot, to come outside with him. I just could not pass this up, so I followed them out. The Alaska pilot told his buddy to hold up the keychain and, looking through it as if it was an HGS, to align the runway marks over the curbs of the road that T'd into the road we were on. The road was like the runway, he explained. He then gave the bizjet pilot HGS-101 using his keychain. I couldn't help chuckling. As he was finishing up, I introduced myself and we all had a great laugh. I complimented the Alaska pilot on his performance and told him I never thought that my little keychain would come in so handy.

Not long after that, I decided to visit the Boeing Store in Seattle. I bought a Boeing shirt and a baseball cap, and then I found some cute "fat airplane" stickers. They had them for virtually all aircraft in production. They gave me a bright idea. I sent one to my brother in Vancouver, BC, who was a great artist, along with a picture of our combiner. I asked if he could come up with a similar design, but for our HGS. A week later, he sent me some artist concepts and they were great! I showed the design to John and Jim and the rest of the A-Team and got a hearty thumbs-up from all to go get some stickers for us.

Going back a little, I need to explain something. One summer day in June, 1987, I was driving to the Seattle airport for my flight to Portland when I heard on the car radio news that an Alaska Airlines B727 had been severely damaged in a fire at a jetway at Anchorage airport. As soon as I made it to my desk, I contacted Tom Johnson at Alaska to inquire about the incident. He told me that an aircraft was being taxied to the jetway by a couple of maintenance workers when one of the wings struck the jetway. The collision ruptured the wing's fuel tanks, setting off a fire that destroyed the aircraft and the jetway. A tug driver saw the fire and was able to tow the burning aircraft away from the terminal, minimizing the fire damage to the building itself. No one was injured.

"Was the cockpit destroyed in the fire?" was my next question for Tom.

"No, I don't think so," he replied. "Why?"

"How can I find out about acquiring the cockpit?" I continued.

"What?" Tom was buffaloed.

I told him it was a chance for Flight Dynamics to acquire our own "simulator." He immediately got it, and provided me with the name of a woman at corporate across the street. I called her immediately.

"You would need to bid on the cockpit if you want me to hold it for you," she told me.

"You mean money?" I asked.

"Yes," was her answer, "or the equivalent."

I mentioned that we had recently signed an agreement with Alaska to buy our HGS for their B727 fleet, and asked if I could bid an HGS computer to "hold the part for us." I told her the value of our computer.

She gave me the green light and I provided her with the part number and details.

I rushed into John's office to tell him what I had just done, thinking he would either give me a pat on the back or kick my ass into tomorrow. To my relief, he was very supportive.

So we began the process of acquiring a "simulator."

It first required our mechanical engineer, Norman Jee, and our flight controls guy, Doug Ford, to fly up to Anchorage to inspect our new purchase. They arrived and found transport to the area where the carcass of the aircraft was being stowed. Luckily the cockpit was relatively intact, although most of the working displays and avionics had been removed. We could live with that.

Then they had to decide how to release the cockpit portion from the rest of the burned-out fuselage. Alaska Airlines had an electric carbide saw that was approved for cutting aluminum. Norm and Doug had to figure out what to keep to convert this mess into a fixed-base simulator. Then, one of the Alaska mechanics used a wide red magic marker to make dotted "cut" lines on the outside of the fuselage where Norm and Doug indicated, while another Alaska mechanic followed along behind, carbide saw screaming. With help from some of the other fascinated Alaska Airlines folks, they managed to load the severed cockpit onto a flatbed truck.

Thank goodness Norm had a plan! The plan involved finding a barge that could take the cockpit from Anchorage harbor down to Portland. With a company credit card and innovation, Norm somehow found what was needed, and finally got the cockpit loaded onto a barge. The trip would take a week.

He and Doug flew back to Portland with lots of pictures so that the folks at our facility could find a good place for our "simulator." A week later, they were at the port in Oregon with another flatbed truck to take our prize to the facility.

But there was another problem!

The optimal route to our new plant would require cutting some power lines to allow passage of the truck hauling our cockpit. Thank goodness for that credit card.

Finally the cockpit arrived at our new facility. It was unloaded into the vacant back area through a large access door. The A-Team gathered around the mess and looked pleased.

About two to three months later, I was asked to come downstairs and see our new fixed-base simulator. I was amazed.

No…I was blown away!

Our folks had cleaned up the rough carbide saw cut, welded together a large metal structure to hold the cockpit, built a set of wooden stairs up to the cockpit door, and installed our HGS. They had also made cardboard cutouts of the instruments Alaska had removed from the burned-out aircraft. One of our talented engineers had even designed and developed a force-feel mechanism to emulate pilot feedback. It actually looked and "flew" like a B727 simulator. It was amazing!

This simulator would come to be a huge asset as we moved forward with new HGS features, symbology changes, and customer demonstrations.

In a discussion with one of my Boeing buddies over lunch one day, I mentioned that we had acquired the cockpit from the ruined B727 Anchorage incident, but it had no real cockpit instruments. He told me about the Boeing Surplus store located down in Kent. There, I found a bunch of instruments that had been listed on a piece of paper by the A-Team as required for the new simulator. Since our parts did not need to be functioning, I got a helluva deal for the lot and brought them back to our plant, where they were installed in our new simulator. It looked much better.

We now had what looked, for all the world, like a fixed-base B727 simulator with our HGS installed.

This was really a fun job!

Boeing and HGS "fat airplane" stickers and HGS keychain

Horizon Dash 8 with HGS installed taxiing in at Portland airport

HGS-2000 Series installed in new deHavilland Dash 8 aircraft

Flight Dynamics "simulator" cut free from Alaska's fire-damaged B727 at Anchorage airport

"Simulator" arriving at the Flight Dynamics facility from the port of Portland

Norman Jee "flying" the sim inside the Flight Dynamics facility

Our fixed-base B727 simulator from the wrecked Alaska B727, with cardboard cut-outs

Lufthansa CityLine Canadair Regional Jets and Horizon Airways Dornier 328s would both be equipped with Flight Dynamics HUDs

Our new Flight Dynamics HGS brochure

Chapter 9

A Big One That Got Away

While all this was going on, the Hughes Aircraft Company down in Southern California had noticed our successes with Alaska Airlines, Horizon Air, and Bombardier, and decided that they wanted to get into the commercial HUD marketplace. They had done some military HUD work in the past. So, near the close of the 1980s, Hughes approached our parent company to explore potential acquisition of our company. After some serious negotiations, the deal was finally signed, and Flight Dynamics became part of the Hughes organization. It was not to last very long.

Mal Currie, Chairman of Hughes Aircraft, assigned one of his key folks, Rick Hilton, to honcho his new commercial HUD business. The HGS marketing team held regular sales get-togethers at the impressive Hughes facility just outside Los Angeles, and Rick regularly attended business, program, and marketing meetings in Oregon. Rick and John Desmond seemed to get along well, and I was impressed once again with John's ability to hold everything together, no matter what was happening to our little company.

On one particular occasion, I needed to join the Flight Dynamics team to visit the Hughes facility in LA. I boarded an Alaska B727 at Seattle airport and managed to score a first-class seat upgrade. I found my seat up front and got settled. Just then Ed Asner, a famous actor on the popular Mary Tyler Moore TV show, boarded the aircraft, and guess what? He was seated right beside me. Shari Belafonte, daughter of Harry, followed immediately behind him and was seated right across the aisle. They were returning from an Amnesty International event in Spokane. After Ed got settled in, he and I introduced ourselves.

As soon as the aircraft door was closed, I brought out some of my B727 fat aircraft stickers and handed one to the flight attendant. She loved it! About five minutes later she returned, grinning, and begged for a couple more. I had plenty—I always traveled with lots of our company stuff. After takeoff, once the seat belt sign was switched off, she came back yet again, and said that the cockpit crew loved the stickers and were asking for a couple more for the pilots. Ed Asner leaned over and asked me what the hell was going on. He said that usually *he* was the main attraction on any airplane and could not figure out why I was so popular. I pulled out one of the stickers and gave him a pedestrian version of HGS-101. He laughed heartily and asked if he could keep the sticker. Of course he could! As he was leaving the airplane after landing, I saw him venture a glance into the cockpit to see our amazing HGS.

By this time, Dick Hansen had become absolutely invaluable in our certification process. Besides being our pilot and HGS simulator demonstrator, he took responsibility for all kinds of tasks inside Flight Dynamics, from symbology and HGS mode development, to FAA meetings, pilot guide preparation, and certification flight testing of the HGS-equipped aircraft. In addition to all these duties, Dick helped with frozen turkey head sales and marketing functions as well.

While I was busy with the Horizon Dash 8 project, and convincing Bombardier to install our HGS on their new aircraft, Dick got heavily involved with selling FedEx on the merits of our system. They became quite interested in it for their twenty-eight older Fokker F27 feeder fleet aircraft operating in Europe. After inspecting the Fokker, the A-Team discovered that the B727 HGS-1000 would require only minor changes to the attachment brackets for the overhead unit and combiner.

Fred Smith, President and CEO of FedEx at the time, was a huge HUD fan. After some key meetings with Ron Wickens, his Chief Engineer, and Jack Finlay in Flight Ops Technical, Fred decided he wanted our HGS on his European fleet of aircraft. Dick asked me along on a couple of the visits. After meetings at their headquarters, we would sometimes meet up with Ron at his favorite restaurant in Memphis, called Folk's Folly. There, he would always order deep-fried pickles for the whole table. I hated pickles, of any kind—a lot—so it was a real challenge to provide the expected compliments on his selection!

The FedEx arrangement was that they would make a down payment on the total contract up front, and pay the remainder when we achieved CATII—a good deal for us. One day in 1989, upon his return from a trip to Memphis, Dick showed up at our office and produced a check for $750,000 from FedEx. We couldn't believe it. And he had done this almost totally alone. As I remember it, our finance department wanted the check returned to FedEx, since we had no spec, no statement of work, no agreement, and no purchase order.

Return the check? Are you kidding?

We didn't return it, and Dick and I laughed ourselves silly over a few congratulatory beers later. OK, maybe it was more than a few!

Later, while Dick and company were doing the certification flight testing in Portland on the FedEx F27, I asked if I could go along on one of their test flights. I had never flown in an F27 before. What a mistake—it was awful! First problem, there were no windows in the back. Since it was used for hauling packages and freight, all the windows were covered with aluminum panels. There were two seats in the back, one for our test equipment operator and the other for an observer. I got the observer seat.

Second problem, the pilot and Dick were looking for lousy weather to analyze the HGS performance. They found some, and the aircraft began heaving and bucking. I told Dick I was going to "toss my cookies." He told me to stand up front between the pilot seats and look outside, and soon the nausea calmed down a bit. But on our return to base I ran for the toilets to give them my cookies.

Sometime later, we achieved the CATII required certification, FedEx paid the rest of the contract, and we delivered the twenty-eight systems plus a couple of spares.

Thanks to Dick, I now had help selling these frozen turkey heads!

But on a later follow-up visit to Memphis, FedEx told Dick that they had decided not to equip the fleet after all, since they were now looking into newer and better-equipped aircraft. So the HGS, still in their original boxes, were stacked up in one of their hangars. Dick asked if we could buy them back, at a big discount of course, and they agreed.

This was to help me a few years later.

After my success on the Dash 8 project for George Bagley at Horizon, I decided in late 1989 to visit him and Perry to see if all was going well. They mentioned that my timing for the visit could not be better, as they were now in discussions with a German aircraft company called Dornier for some newly-developed thirty-two-passenger DO328 turboprop aircraft. They told me I should expect a call from one of Dornier's salesmen to ask about our HGS. Horizon would be the launch customer for this new aircraft, and they had told the European OEM, "No HGS, no deal!"

This was just getting better and better every day!

Sure enough, a couple of days later a call came across to my desk from the US-based Dornier salesman, Randy Becker, who asked me to give him the telephone version of HGS-101. After a lengthy chat, he told me that I would need to go to Friedrichshafen in Germany to provide a presentation to Dornier's technical crew there. Where the hell was Friedrichshafen? He said that Wolfgang Adam would be my contact at the facility.

Randy Foster, one of our customer support folks, and Norman Jee, who had done such a good job on our simulator project, agreed to go with me to address any of the really "down in the weeds" technical stuff, like required wiring, systems interfaces, and necessary structural modifications, that would surely come up during the presentation and discussions.

We flew to Zurich, Switzerland, which was the closest international airport to the German aircraft manufacturer's facility. It was the start of the first Gulf War, and the Swissair flight was virtually empty. After landing, we rented a car and drove to Friedrichshafen. It was a long drive and included a chilly ferry ride across Lake Constance in the middle of winter.

When we arrived at the Dornier plant the next day, a friendly Wolfgang met us in the lobby and escorted us to a conference room. There we met the technical team, and gave a presentation on what we had done for Horizon's Dash 8 aircraft.

After a full day of meetings, we felt things were going extremely well. We had also met one of their most senior executives, Jack Pelton, who was very supportive of the HGS and told us he wanted to add our system to their newest concept aircraft, the larger seventy-two-passenger DO728. Unfortunately, this aircraft never saw the light of day. (Quite a bit later, Jack became VP of Engineering, and later still, he was promoted to president of the Cessna Aircraft Company back in the US.)

Before we departed, Wolfgang took us on a tour of their manufacturing facility, where they were milling whole aircraft wings from large slabs of aluminum. Impressive!

The deal we eventually struck was that Dornier would install the HGS attachment hard points, the system wiring, and all aircraft-related modifications prior to delivery, and we would ship the HGS units directly to Horizon's Operations Center in Portland. There, we would install our system and complete a duplicate STC certification task like we had done for their Dash 8s.

A few weeks later, Wolfgang called me to announce that he wanted to visit our facility on his scheduled trip to meet with Horizon in Portland. I greeted him on his arrival in our lobby, gave him a tour of the facility, and introduced him to our senior management. The girls at our plant went nuts over the "good-looking Hollywood dude" from Europe. We had to help him fend them off, and catch any that fainted.

The Dornier DO328 HGS project was a success, but the aircraft did not fare as well. Unfortunately, much later, one of their early delivery aircraft had an incident in the winter flying out of Portland. Snow and slush from the runway splashed up into the wheel wells and, after takeoff, froze the wheel well doors shut. The pilots had to do some "creative maneuvering" to get the wheels released for landing. The main undercarriage for the DO328 was attached directly to the bottom of the fuselage and not to the engines, like the Dash 8, making it susceptible to this issue.

Horizon cancelled the contract with Dornier.

After Delta acquired Western Airlines, Larry Hecker went to work for the FAA in Washington, DC. He called one day and gave me contact info for a friend of his at Northwest Airlines in Minneapolis. Northwest had a larger fleet of about sixty B727s, as well as some other aircraft, including wide-body B747s and DC10s. Larry had briefed his friend on the value of considering our HGS for their fleet, and the friend was interested. His name was Captain Dave Haapala.

So one day in 1989 I called Dave on the telephone. He told me that Larry had done a real job on him. And he was ready to listen. I took him through the telephone version of HGS-101, which I was now getting very good at, plus a lot of the Alaska campaign details, including the famous Twelve Days of Christmas in '85 and our ultimate justification for the HGS. Dave clearly remembered the event. Northwest had also been hit, although not as bad as Alaska, since they'd had fewer flights into and out of Seattle during the fog episode. He told me to mail him some HGS info.

I sent him our HGS brochure, some newer aviation magazine articles detailing the Alaska decision, and some articles about Horizon's Dash 8 commitment and the support of the OEM, Bombardier. (Bombardier was now also offering the HGS on their brand new fifty-passenger regional jet, the CRJ.) I mentioned that our system had been selected by Lufthansa Cityline, Tyrolean Airways, and a few others, and I threw in some B727 HGS stickers.

After receiving my package, Dave called to suggest that I visit Minneapolis for an HGS presentation. He would invite their key folks to "Captain Sunshine's magic HGS show."

I flew to Minneapolis and met Dave for the first time at the Northwest Flight Ops building. We were to become good friends as well as business colleagues. Dave was a specialist in their Flight Operations group and was involved in assessing new technology for Northwest. He was in the perfect position to advocate for our system.

Dave introduced me to some of his colleagues: Jim Magus, Duane Edelman, Bob Buley, and a bunch of other folks who would eventually become involved in the project. We had lengthy discussions with Don Nelson, their VP Flight Ops, about the payback.

Northwest management were very astute about their airline weather disruption costs. Don and Dave decided that we would need to develop a whole new payback model directed specifically at our HGS, and they had a lot of great ideas on how this could be done.

John and Jim were supportive of the new sales target. As time went by, I found myself flying to Minneapolis almost every other week. On one of these trips, Dave met me at the entrance to the Northwest Flight Ops building and invited me for a drive. I hopped into his car and we drove around the end of the airport toward the landing zone of one of the runways. We chatted about the HGS project and the collection of Northwest weather disruption information until Dave stopped the car and turned off the engine.

"What are we waiting here for?" I asked him.

"We are here to witness a very important event, Phil," he explained. "You see, Northwest is the first airline in the US to order the brand-new Airbus A320 aircraft".

"Wow! I've been reading about it but I've never seen one yet," I replied enthusiastically.

"Keep watching," Dave continued, "and you'll see one any minute now. Our first delivery is scheduled to arrive in about five minutes."

Sure enough, as we watched, the new Airbus aircraft gracefully flew over the airport fence and landed right in front of us. It was beautiful. The A320 was the first Fly-By-Wire (FBW) commercial aircraft in the world, meaning that it used electronics instead of flight control cables. I was thrilled to witness the event.

Dave started up his car and we drove over to the hangar, where we were allowed into the Airbus to take a look. It smelled new, like a new car. I had never seen side stick controllers instead of control yokes in the cockpit of a commercial aircraft before. I was impressed.

Back at Dave's office, he took me down the hallway and showed me an unoccupied office. He pointed beside the door, where my name was on a small plastic sign.

"What this?" I asked, surprised.

"This is your new office," Dave said. "You'll be visiting us so much, we thought it only right for you to have your own office here." John, Dick, and Jim were quite impressed when I told them later.

One day on a subsequent visit to Northwest, Dick, John, and I were staying at the Holiday Inn near the airport. It was the middle of winter and it was cold…damn cold. The temperature outside was minus 50°F, with wind chill to minus 70°F. The morning of our meeting we had breakfast at the hotel. John asked if I would be a good sport and go get our rental car from the outside parking area to warm it up. I was Canadian after all, and therefore used to real cold. The car barely started. I drove slowly over to the hotel lobby. It had a protective overhang, but heat from the lobby had melted some of the ice, which had trickled down to the roadway in front of the main entryway and frozen again. As I drove under the canopy, I touched the brakes and nothing happened. The car continued to slide…right into the back of the car in front of me. I had been going very slowly, but the contacting cars made a crashing sound like a wineglass dropped onto a concrete patio. I got out to look. My front bumper, which was plastic, had shattered into a million pieces, the largest no bigger than a quarter. The other car did not have a scratch.

Dick came out, took a look, and, laughing his ass off, went back inside to ask the concierge for a plastic garbage bag. We swept all the pieces into the bag and threw it in the trunk. After a great meeting with the Northwest folks, we returned the rental car to Hertz. The agent came out to check the car over.

"Where's the front bumper?" he asked.

"In the trunk," I replied.

He opened the trunk, looked inside, and said "Nope, not there."

"In the garbage bag," I said.

He opened the bag and almost fell over laughing. I told him the story and he said that the exact same thing had happened to someone else's rental the day before. Damn new cars…

The hard work on our payback effort started. I needed a computer modeling expert. John assigned one of our young employees, named Tom Geiger, who happened to be the son of John Geiger, the original president of Flight Dynamics. Tom and I became good traveling companions, flying out to Minneapolis on a regular basis.

For his first trip to Minneapolis, I had warned Tom to wear a very warm jacket since it was still mid-winter and really cold. I had flown in a couple of days earlier for some discussions with Dave and offered to pick Tom up at the airport on his arrival. As we met at the luggage retrieval area, I burst out laughing at Tom's idea of a warm jacket. He was wearing an Oregon raincoat!

"What's so funny?" He looked at me, quite puzzled.

"Just wait till we go outside," I responded. "You'll see."

As the airport's automatic doors opened up, Tom literally gasped.

"Holy crap," he chattered.

On the way to our hotel to check him in, we popped into a Sears store where he was able to pick up a real winter coat before he froze to death.

On another occasion, back in the Pacific Northwest, Tom flew up from Portland to Everett—Horizon actually offered that flight—to continue the Northwest work with me, because I had broken my collarbone dirt-biking. We both had a good laugh about that one.

On one of the Northwest visits to Minneapolis, we had been working on getting data into Geiger's computer for three to four days straight. Dave arrived on our last morning and informed us that we were going to take a break from our hard work that day. He loaded us into his car and drove over to the Minnesota State Fair. We spent most of the day walking around the huge event and eating corndogs and other fair specialties. I think Tom and I even went on a couple of fair rides while we were there.

Northwest had a lot of good data on poor visibility disruptions. Combining it with my previous data from Alaska Airlines, Dave and I came up with some good ideas about how to build his new disruption model, which was gradually getting worked into the computer by Tom. Every other week, back in Portland, Tom and I would update John, Jim, and Dick on the model's development, and progress at the airline. Everyone was happy.

Then another interesting thing happened.

As Dave, Tom, and I pored over the Northwest historical weather disruption analysis and data, it became quite clear that the airline hits taken during a foggy event by a wide-body aircraft such as a B747 or a DC10 were far more serious and costly than for the smaller aircraft. The downstream effects were particularly ugly.

That result should not have been a surprise.

The larger aircraft carried 250 to 350 passengers each over international routes that yielded high revenues for the airline. Often during a weather disruption, the carrier had to put all the passengers up in hotels, or pay other carriers to complete their cancelled flights or missed connections. And there were other costs, like the amount of holding and diversion fuel needed for a three- or four-engine wide-body aircraft. Dave decided we should expand our study to include HGS for the entire Northwest legacy fleet of aircraft. This even included a large fleet of older DC9s. I was worried that we were getting over-extended, and the cost of the HGS program was sky-rocketing, especially since many of the target aircraft needed the IRS and the DADCs. I tried unsuccessfully to convince Dave to reduce the scope of our effort back to the B727 fleet only—but the cat was out of the bag, and there was simply no way to stuff him back in.

After we returned to our plant, I told the team about my rapidly rising concerns with the expanding scope of the Northwest HGS project, and the sheer size, cost, and complexity of our tasks. It wasn't just the quantity of aircraft, but all the various configurations of different aircraft types—Northwest had acquired aircraft from many different sources. It was a bit of a mess. Actually, it was a huge mess!

But with Flight Dynamics management encouragement, we decided to truck on.

On a subsequent trip to Northwest with Tom Geiger, we heard that Northwest had put together a spec of sorts for a HUD for their aircraft. We needed to know if this Flight Ops Technical Spec was written with our HGS in mind. We knew they had had some discussions with Sextant-Avionique, a French HUD manufacturer. If we could just take a look, we'd know if we met their spec and had a high likelihood of being selected for the program.

We were working late one evening at the Flight ops building, maybe 7 pm or so, and all of the admin folks had left for the day. We were holed up in our new quarters near Haapala's office. Dave wasn't in his office, but we didn't know where he was.

I looked at Tom. "Hey, I need to make a copy of something—follow me." I quickly grabbed the spec off Dave's desk and took it to the copier room. "Watch the hallway and let me know if anyone is coming," I instructed Tom.

I had a hell of a time getting the copier's automatic feeder to work—gotta love those old copiers—and it seemed like an eternity to both of us. After a lot of fidgeting, fiddling, and swearing, I finally got the damn thing going properly. I grabbed our copy and we raced back to return the original document to Dave's desk, then back to our office to get packed up for our early morning return flight to Portland the next day. Seconds after we returned to our room, Dave came in and suggested a place for dinner—whew! But I got the distinct impression that a smiling Dave had wanted us to find the document.

The subsequent A-Team review of the document reassured us that we were the front runner.

After we'd been working the new payback model for about six months, Dave told me that his team was ready to take the project to executive management, and eventually to the Northwest Board, and he had the events already scheduled. John, Jim, and Dick were elated once again, since this had been such a good sign at Alaska Airlines.

Then I got the dreaded phone call. It was June, 1989.

"Phil, we have to postpone our Executive Presentation," Dave announced.

"Why," I asked, becoming quite concerned.

"Can't say just yet," he replied. No! Not again...please, not again!

"Geez, Dave. How long?" I enquired.

"Indefinitely," was his response.

He had been working his executives in Flight Ops and they had been working their bosses all the way up the food chain to Bill Slattery, their President, and Steven Rothmeier, the airline's CEO. But it would all be to no avail.

Finally, about a week later, Dave told me that Northwest was being acquired—not by another airline, but by the Checchi Brothers, West Coast bankers and investment specialists who had insisted on freezing all capital assets until after the acquisition was completed.

God help us, it was happening again...

The end result was really bad for us. After completion of the acquisition, the brothers decided to sell off a lot of the Northwest assets and do lease-backs of buildings, aircraft, and other assets. I knew without a doubt that this would be the end of our HGS project at Northwest, and it was. It was a real blow.

But we did gain a really good HGS payback model out of the whole effort. Once again, Captain Sunshine tried to see the brighter side of things.

In my office later, I learned from John that we had hired a Boeing specialist, George Kanellis, whose expertise included working payback models for Boeing's airline customers. After meeting George, Tom and I showed him our extensive and detailed Northwest HGS payback model. He was impressed, but suggested that we needed to use the format and language understood by airline finance folks. OK. He reworked the entire model to his airline standards, and even though I could not understand it as well, he assured me that airline finance execs would get it. George would later write a chapter for a book on airline finance that was based on our flight disruption analysis work for the HGS.

Our marketing team was now meeting regularly for lunch at Wu's Chinese restaurant, across the highway from our facility in Tualatin. We would share a few Asian dishes and chat about the HGS and the status of current sales campaigns. I was definitely down after the Northwest loss and tried hard not to bring the rest of the team down too. After all…I was Captain Sunshine.

As we were returning to the office from our Wu's lunch one day, our receptionist intercepted me, handed me a piece of paper with a fellow's name and phone number on it, and told me that I should call him. The number was for General Motors in Detroit. What the…?

I dialed and asked for the name on the paper. He was in the marketing department at the auto company, and immediately starting asking questions about our HUD technology. When I enquired why, he said that they were looking at putting HUDs in cars, and he and a couple of his folks wanted to come to our place to chat. I mentioned this to John, who was quite skeptical at first, but agreed to at least meet up to see what they wanted.

A few days later, the GM team showed up at our facility, along with a fellow from Pittsburg Plate Glass (PPG), the company that built all their auto windshields. After settling in to our conference room and introducing themselves, they explained to us that they had been thinking about HUDs for automobiles for some time now. John was incredulous.

"Why the hell would you want to put a HUD on a car?" was John's main question. He was thinking about financial justification, but maybe it was all about safety.

"Why, to sell cars, of course," was the reply. That simple!

They had brought along the PPG fellow since, over the phone, I had told them about our HGS combiner holographic sandwich, and they wanted him to chat with Bob Wood to learn more about how to do this—was it even possible for a car windshield?

They then explained that they had designated a Camaro Berlinetta as a test vehicle, and could deliver it to our place to allow us to work with PPG to get a small HUD into the Camaro. How much would we want to do this? After a bit of dickering, we arrived at three hundred thousand dollars. Deal—the car would be at Flight Dynamics in two weeks.

For the next while I thought little of it, since I was up to my eyeballs with our developing strategy for Boeing. Then, some weeks later, Bob Wood told me to come downstairs. Lo and behold, there was the Camaro, and it had a HUD in it, with detailed driver symbology. It looked fantastic. PPG had embedded the hologram into the Berlinetta windshield and shipped it to our place, and Bob's guys had been able to remove the original and replace it with the hologram version.

At a subsequent company marketing meeting, John suggested we bring the Camaro to Seattle to show Boeing that it was possible to embed the hologram right into an aircraft windshield, eliminating the need for a separate combiner. I was nominated for the task.

On my day of departure for Seattle that morning in 1990, Bob told me that the hologram windshield was quite expensive, and he was afraid of gravel chips on the long drive up. So his guys would put the original windshield back in for the trip to Seattle, and stow the holographic version in the trunk. He wanted me to watch his guys do it so that after arriving in Seattle I could switch back to the hologram windshield for the Boeing demonstration drives.

I had already prepared some of our Boeing colleagues and acquaintances for our visit. Most of them thought that a Camaro with a HUD installed was a hoot, and they were really looking forward to seeing and driving it.

Before my departure, Bob again cautioned me to be careful with the holographic windshield…we only had one.

I hopped into the Camaro and headed out. About fifteen minutes into the drive, I thought I would turn on the HUD to see what it looked like, even though the car had the wrong windshield. I was shocked! The symbology was bright and crystal clear and, for a minute, I thought the guys had forgotten to exchange windshields. I pulled over at the closest 7-Eleven and called back to the plant. After getting a hold of Bob, I told him that the symbology looked just fine…maybe even better without the special windshield. He demanded that I bring the car right back for him to take a look. They had never tried turning on the HUD with the standard windshield.

The whole group gathered around the car and was amazed. I had been right—the HUD symbology looked great with the regular windshield. Bob immediately understood why, and explained that, because of the significant rake angle of the car's windshield, we were getting great reflectivity from the HUD's projector. Bob pointed out how annoying it was when you put a piece of white paper on the dashboard of your car, because the reflection was so bright. It looked like we did not need a specialized windshield at all.

We left the original windshield in and I headed to Seattle, where I met the PPG rep. The two of us conducted twenty to twenty-five driving demonstrations for Boeing engineers and flight test folks over the next few days. Everyone was amazed.

Sometime later, GM realized that they also owned Hughes, our parent company, which had built many military HUDs in the past, and had the ability to build HUDs for their cars. We were actually quite pleased about this development, because they had told us two things that worried us a lot: first, they would need about 20,000 HUDs a year and second, they could not cost more than $300 each.

Better to let Hughes do it.

Meanwhile, Dick was busy again doing the frozen turkey head thing. He was getting good at it!

Back in the late 80s, we had supplied an HGS-1000 to the Lockheed

Aircraft Company for installation in their HTTB, or High Technology Test Bed aircraft. It was a modified L100, a commercial variant of the popular C130 military transport aircraft. Lockheed had discovered, after looking at some of our drawings, that our B727 HGS-1000 would fit the HTTB aircraft with minor bracketry changes. This test aircraft was in support of a planned new C130 variant called the "J" model. The test program gave Lockheed a great opportunity to investigate our HGS for this new aircraft model. Dick had even flown the HTTB. (Later, in February of '93, the aircraft crashed on a test flight out of Marietta, Georgia. Thankfully the cause had nothing to do with our HGS.)

Lockheed had now moved a long way toward the new aircraft design and development, and wanted to get into serious discussions about our HGS for the "J." They wanted the aircraft to include dual HGS, and these would be basic equipment—that is, standard on all aircraft delivered from their factory. After a lot of system development work by our A-Team, Flight Dynamics delivered the first systems to Lockheed for installation in their J-model simulator. This was a means to ensure that all was good with the symbology, optics, systems interface, and pilot impressions.

Lockheed was extremely pleased with the results. At the Paris Airshow in 1995, they allowed Flight Dynamics to announce Lockheed's selection of our HGS for the new plane model.

We delivered a few systems to support their aircraft flight test program, which would begin the following year. The company was very satisfied with the performance of our systems, and told us that the aircraft had been selected by the USAF, Britain, and Australia, for a total of around three hundred aircraft, including options.

Dick had done it again.

More firsts—a military OEM and a dual HGS fit!

A Northwest Airlines B727 taxiing for takeoff

FedEx F27 undergoing HGS alignment at Portland airport (I will toss my cookies)

Norman Jee (left) and the Dornier team in Germany to install the HGS in a DO328 for Horizon

Randy Foster and me on the ferry across Lake Constance

GM delivers their Camaro Berlinetta to Flight Dynamics

Our first experimental auto HUD in the Camaro

Chapter 10

Intro to The Big B

Ever since that first dismal meeting years ago with Jack Britt at Boeing, I knew that somehow, sometime, some way, we would need support from this major OEM, but I was not at all sure how we would ever achieve that.

At our regular sales meetings with John Desmond every Monday, I was asked "How are things with Boeing?" I continued to explain that Boeing would be a long slog, and that progress was expected to be slow. However, I decided after our successes with Horizon, Bombardier, and Lockheed, that I should now spend a little more time trying to move this aviation giant in our direction.

Around late 1987 I spoke to a number of folks I knew at Boeing, seeking guidance. Two names kept coming up: Chris Longridge and Carl Lund. Chris was VP of Sales & Marketing for Boeing, and Carl was a development engineer for the B737.

I was able to get Chris's contact information through some marketing folks I knew. After a telephone "executive version" of the HGS brief, he agreed to a meeting. John Desmond decided that he should tag along. We flew up to Seattle and when we arrived at Boeing we discovered that Chris had invited Carl to join the meeting.

It was to be a very good meeting.

Chris had read some complimentary articles about our HGS in the media. He had also heard positive comments from some of their airline customers through his sales staff, as well as from his customer engineering and other support groups. Apparently airline customers were actually asking Boeing for the company position on the HGS, so Chris was receptive from the get go, and became even more supportive after our meeting.

At the meeting, we gave Chris and Carl the full HGS-101 course. Like others I had met at Boeing, Carl was a little skeptical at first, but I convinced him to take a look at it in the simulator. Following a great simulator session with Dick Hansen, who was now an absolute master at HGS demonstrations in the M-Cab, Carl came out a changed man. Like me, Carl was not a pilot, but with his strong technical aviation and avionics background, he immediately understood the benefits of the system. He was more than a little impressed, and immediately we began to strategize how to get the attention of Boeing's upper management.

Chris and Carl worked wonders from the inside! Over the next year, they introduced me to many others at Boeing who would become key to our progress and eventual success there. Pete Rumsey was Carl's immediate boss. Jim Von Der Linn was in engineering support to marketing. Steve Henderson and Dennis Morden were in marketing and reported directly to Chris Longridge.

Then there was Kenny Higgins, VP of Flight Operations. Kenny had no formal flight training—amazingly, he had taught himself to fly using the Boeing simulators, and was now qualified to fly all airplanes in production at Boeing. Dick and I met with him, gave him the HGS brief, and invited him to come and fly the M-Cab. Dick did a great job, as usual, and Kenny became extremely supportive after flying all kinds of tricky and sporty approaches, including a simulation of a recent serious incident involving a Delta Lockheed L1011.

The Delta plane had crashed at DFW in August, 1985, as a result of what aviation specialists were now calling wind shear. This occurred when an aircraft flew through a strong weather-related downburst of air that spurled out in all directions as it hit the ground—think of an upside-down mushroom. At altitude, this is not a problem, but if it occurs on approach for landing, the result can be disastrous. The aircraft's airspeed increases as it encounters the mushroom's outflow, so the pilot naturally reduces power in an effort to maintain the correct airspeed. But suddenly the aircraft meets the middle, violent downward push of the air, which forces the aircraft down and closer to the ground. Now the plane encounters the mushroom's outflow as a tailwind, which reduces the airspeed even further, and therefore the aircraft's lift. Without instant recognition of what is happening, the result can be fatal.

Having stayed abreast of the investigation results, the A-Team's Doug Ford had worked out algorithms that were added to the HGS software by Ken Zimmerman. The modifications detected and confirmed wind shear conditions and provided escape guidance out of such an event. Dick had tested it in the Boeing M-Cab many times and it worked every time. Once again, I was amazed at the A-Team's technical skill. The upgraded system was named the HGS-1000WS.

We soon had full support from Pete Rumsey, Jim Von Der Linn, and the marketing team at Boeing. But all of us knew we would eventually need the OK from Alan Mulally and his boss, Jim Johnson, if we were to be successful. Alan was then the VP of Engineering in Renton (he later became the President of the Boeing Commercial Airplane Company). Kenny had promised to tell Alan about his simulator experience and to vouch for our system.

I had met Alan once before while working for Simmonds, and also knew his secretary. I mentioned to John Desmond that we would need to meet with Alan in person to get to any next steps at Boeing, and he agreed. We had heard from some of our Boeing colleagues that Alan was not an enthusiastic fan of the HGS and had shut down some activities inside that were leaning in our favor.

In preparation for a possible meeting with Alan, I asked the Boeing team what the main issues were regarding our HGS. They explained that there were quite a few serious problems.

First, our system had "deep roots." This meant that the HGS was tapping into many different sensors and systems throughout the aircraft.

Second, Boeing had not written any detailed specification for the HGS—Flight Dynamics had designed and developed the system completely on our own, without any guidance or inputs from this OEM.

Third, it flew in the face of Boeing's flying philosophy that each pilot should have all flight information.

And fourth, Boeing's approach was to use autoland to handle poor visibility landings (as Jack Brit had explained to me previously).

The team told me that many people at Boeing saw the HGS as a sort of threat. Over the years, the company had put a lot of effort and resources into the design and development of the aircraft's autoland system, as well as the cockpit displays, and our HGS would put these systems in second place.

I finally began to understand.

Taking a chance, I called Alan's secretary one day and asked if we could schedule a meeting with him for the following week. She asked if I had coordinated with Alan and I lied, "Yes!"

I briefed John on what I knew about Alan—highly technical, get to the meat, don't beat around the bush, and if you don't know the answer to a question, say so and promise to find out and get back to him.

John and I flew up to Seattle in his own plane, a twin-engine Beech Baron. After renting a car at Boeing Field, we arrived at Alan's office. His secretary greeted us and went in to announce our arrival. Alan came out and shook my hand, and I introduced him to John. He invited us into his office. On the way in, he leaned over, put his arm on my shoulder and said quietly, "that was not the right way to get in to see me." I was mortified. But we rapidly got into an extensive discussion on the HGS, the technology, and its benefits, and I forgot my embarrassment. Alan had obviously heard about the product, certainly from his staff and possibly from the industry magazine articles that were appearing on an ever-increasing basis. John did an absolutely superb job of answering some very difficult technical questions from our skeptical host. Suddenly, Alan leaned over his desk towards John.

"So, in these wind shear conditions, why do you think the aircraft are getting into trouble?" he asked.

John looked at me nervously. I shrugged and indicated he should keep going. There was silence as he tried to formulate the best response to Alan's question. Just when it was getting awkward, John leaned back over Alan's desk.

"Because," he said emphatically, "the pilots don't know where their aircraft is going."

"You are exactly right, John," was Alan's response.

John was elated, as was I. The rest of the meeting went very well. As we were preparing to leave, I offered an M-Cab simulator session for Alan to see our "baby." He accepted and told me to coordinate with his secretary…but the right way this time.

Three weeks later, John and Dick were flying to Seattle for our scheduled M-Cab session with Alan. I drove, since I had other tasks at Boeing after the session. John insisted we arrive the night before, in case we faced a repeat of our earlier Bruce Caulkins fog episode fiasco. We all stayed at the Holiday Inn in Renton.

That evening, over dinner, we discussed the importance of our pending event with Alan, and thought about any possible objections he might bring up and how to address them.

We awoke the next morning to snow! Lots of snow. In Seattle! John was more than a little upset.

I had been introduced previously to Dale Ranz, a Boeing technical pilot. Alan had asked me to invite Dale to the HGS M-Cab session. I called over to the M-Cab, where Dale was already waiting, since he lived close by and had managed to get there, despite the snow. He suggested that we cancel since, most likely, Alan would not make it. He lived in Woodinville, about thirty miles north. I asked for Alan's home number, not something easily come by, but Dale gave it to me. I called Alan. He told me his driveway was four to six inches deep in snow and we would have to postpone. I offered to come pick him up in my Chevy S-10 Blazer, a four-wheel-drive vehicle, but while he was impressed with my enthusiastic offer, he still declined.

We scheduled another M-Cab session for him not too long afterward, and while waiting for things at the simulator to get set up, we chatted and got better acquainted. Alan had many questions about the Alaska campaign, which seemed to fascinate him. He also knew that Phil Condit, another Boeing senior executive at the time, and Alan's ultimate boss up the food chain, had flown the HGS and liked it. And Alan had heard good things about our system from Kenny Higgins and Chris Longridge, both of whom he respected.

In the M-cab, Alan sat in the pilot's position, Dale Rantz was in Dick's co-pilot seat, and John, Dick, and I sat behind. The simulator was started and Alan flew the HGS in utter silence. At landing, he simply said, "Again."

He flew quite a few landings, including the Delta L1011 wind shear scenario, without saying much, and I began to have serious worries about his experience and first impressions. Finally, after another successful landing through the wind shear event, Dale stopped the simulator. Alan turned in his seat and looked at all of us. We held our breath…

"You know what?" he announced. "This is the way it's supposed to be."

Holy cow! Did I just hear that? Alan was coming around.

As he was leaving the M-Cab, Alan gently reminded me that Boeing was a very big company with a lot of inertia, that they could not move fast, and that this was a radical departure from the way they designed and developed aircraft. He told me not to expect any quick results, and that I would need the patience of Job. He also mentioned that a good customer asking for the HGS on a significant new airplane order would certainly help our cause.

With continuing efforts by Carl, Pete, and the team at Boeing, there was now mounting interest in the possibility that the HGS's certification for CATIIIa might allow CATIIIb (300 RVR) landing if used in conjunction with the aircraft's CATIIIa autoland system. In other words, could two CATIIIa systems equal CATIIIb?

Boeing's primary competition, Airbus, was touting the CATIIIb autoland benefits on their A320, over the B737's more limited capability. And they were making hay with many potential airline customers in Europe, where the lower minimums were a crucial component of operational efficiency. So, in late 1987, Jim Johnson, the GM of the Renton Division at that time, and Alan's immediate boss, decided to "pre-implement" Flight Dynamics to investigate the possibility of using our HGS to achieve CATIIIb on the B737. This meant that our little company would be funded by Boeing to investigate the HGS-autoland combination, or hybrid, to see if it was even possible, or certifiable. This was exceptionally great news at our facility, and for the A-Team. Boeing was paying us, our buddy Carl was running with the program, and we knew we could count on his full support.

With some encouragement from the Boeing team and our friends at the FAA, we became convinced that the hybrid concept was viable, but we would need a lot of help from Boeing to accomplish it. The concept had the autoland conducting the approach, with our independent HGS doing the monitoring. If anything looked out of whack, the pilot could either take over on our HGS to do a go-around, or continue to land if the minima were satisfactory.

In the early summer of 1988, we got even more excited when Carl told us that his boss, Pete Rumsey, had been instructed to do a market survey of airlines to gauge interest in the HGS system. Targets were in both the US and Europe—mainly the UK, Scandinavia, and some mainland European carriers. After visiting the European airline targets, we were told to hit up United, Delta, Northwest, and American on our way home.

There were to be two teams. I was on Team One with Jim Von Der Linn and Pete Rumsey. Dick, Carl Lund, and Jim Gooden were on Team Two. There were dual objectives for the trip. First, get a handle on potential HGS interest itself. And second, explore the new hybrid HGS-autoland possibility for CATIIIb operations. Could the airlines accept this approach?

To ensure that our story was consistent, a lot of coordination was required, especially for our presentation and our answers to any airline questions. We had a number of rehearsals.

Over the two-week trip, our teams saw more than fifteen airlines in Scandinavia and Europe, operating a total of over two hundred B737s (-300s, -400s, and -500s). With only one exception, all of them were extremely supportive of the HGS product and the hybrid concept. (We had flown representatives from most of them in the M-Cab at one time or another.) The airlines mentioned a wide variety of ways in which the HGS could benefit them, and we captured them all.

The teams were to meet up in London at the Sheraton Knightsbridge Hotel about half way through the trip. To ensure that there were no complaints of bias in our results, we would shuffle members before continuing. At the mid-trip get-together, we learned that the second team had had similar positive responses so far.

All the members of team two made it to the UK except Jim Gooden, who had fallen seriously ill after catching some kind of Scandinavian bug. He ended up spending over a week at a hospital in Stockholm, but was able to meet up with us for our return flight back to the US. Jim told us that Sweden's medical service was great, and he had been expecting a hefty charge for his stay. However, under the socialized medical system in that part of the world, he walked out with a big zero on his bill.

The trip went well until we got to United Airlines.

Jim Johnson had some good connections at this airline and decided he would contact them directly to ask how they felt about the HGS.

There were two problems.

First, United had apparently looked at retrofitting the HGS to a large fleet of B727s shortly after the Alaska Airlines HGS announcement. At the same time however, the United aircraft were undergoing hush-kitting, an engine modification to meet new noise abatement regulations. United had decided against the HGS because of the additional costs of the IRS and the DADCs, on top of the costs of the hush-kitting program. Besides, they were looking at buying replacement aircraft in the next four to five years, and these newer aircraft would all have the CATIIIa autoland capability, which they felt eliminated their need for the HGS. They saw no serious benefit in the US for CATIIIb.

Second problem. No one from Flight Dynamics had been keeping United "warm." We had not kept them up to date on some of the other reasons for choosing the HGS. Even the selection of HGS by Alaska, Horizon, and Bombardier didn't sway them.

So when Jim Johnson called his United connections to gauge their interest in the HGS, he ran into a brick wall. He felt that the negative inputs from United outweighed all the positive responses the two teams had received from the European and Scandinavian airlines, and he vetoed any further discussion of installing the HGS in Boeing planes. We were all surprised and devastated.

Not long after our return from Europe, British Airways, the only naysayer on our European tour, announced their intention to purchase fifty B737-400s, with options on more. During some earlier research I had done, I'd learned that BA was actually one of the original co-developers of autoland, on a BAC111 aircraft many years earlier. For their new order, they were demanding a CATIIIb autoland system on the Boeing aircraft.

This event had an awful outcome for Flight Dynamics. Jim Johnson terminated the Boeing pre-implementation project, cutting off our funding, and instead investigated what it would take to upgrade the B737 autoland to the CATIIIb capability. When British Airways was told that the Boeing estimated price per aircraft was 1.5 million dollars for a CATIIIb autoland, they declined. Later, they decided to order the Airbus A320 instead of the B737-400. It was bad news all around.

But it turned out that HGS at Boeing was not dead after all. Two inside champions were to continue to support Carl's efforts in the early 1990s: Jim Von Der Linn and Chris Longridge.

Jim, a strong HGS supporter, risked his career at this OEM to get the HGS offered as a Boeing factory option (we'll get to that story later).

Chris had become a huge fan almost immediately after that first meeting with John and me, and went to great lengths to try to convert some of the anti-HUD folks in engineering and upper management at Boeing.

Chris informed Boeing management that airline customers were starting to become more than a little interested in the HGS. He also convinced some of the key customer engineers and their executives at Boeing to get into the simulator and fly it. He asked Dennis Morden and Steve Henderson to assist me with introductions to some of the Boeing customer engineers and to set up simulator sessions for them. Dick and I spent more and more time with this group. We quickly realized that they were Boeing's primary interface to airline customers, and we wanted positive and supportive responses to any airline HGS inquiries.

During a subsequent meeting with Chris, he told me he had someone else he wanted me to meet. His name was Borge Boeskov, Director of Boeing Sales for Scandinavia.

I liked Borge immediately. He knew about the HGS, had "flown" it in the M-Cab, and had spoken positively to his Scandinavian airline customers about it. That explained some of the very positive inputs we had received in that part of the world on our investigative trip with Boeing.

But Borge told me he had a whole new deal cooking. Cautiously, he explained that Boeing and GE were forming a joint venture, and were calling it Boeing Business Jets, or BBJ.

"Phil, our plan is to convert the Boeing B737NG into a global business jet," he announced proudly.

"You're kidding!" I retorted, almost laughing. "Borge, who in the world would buy a B737 as a bizjet?"

But I could tell that the smiling Borge was dead serious. He told me he would very much like our HGS as a customer option for his new BBJ aircraft.

A couple of days later I got a call from Borge with a new name: Mike Hewett. Mike was the Boeing Chief Test Pilot for the B737, but was soon to become the Chief Pilot for the new BBJ organization. He wanted to have a look at the HGS in the simulator before he would endorse it. Dick and I arranged it and met Mike at the M-Cab on the prearranged day and time. As we approached the cab, our Boeing simulator operator cautioned us that they had been doing some required repairs to the simulator, and had removed the padded cover on our HGS overhead projection unit.

To get motion on the M-Cab simulator, which always made the experience more real, you had to push a button on the panel above the pilot seat. Every once in a while this would cause the whole simulator to lurch forward, and we had always cautioned any new pilots about this potential issue. But for some reason we assumed Mike knew about it.

When Mike reached up and pushed the "motion on" button, the cab lurched violently. Without his seatbelt fastened, he was thrown up and collided head-on with our now exposed, and very hard, projection unit. Suddenly, from my seat in the back, I noticed blood…a lot of blood! He was bleeding profusely from a bad cut to his forehead. I raced out and got some paper towels from our operator, then Dick and I helped Mike out of the M-Cab and found our way to the Boeing medical office for some treatment.

We were worried that this negative experience would come to haunt us in the future. But sometime later, Mike actually joked about it. "You can always tell an HGS pilot by the scars on his forehead," he said, laughing.

One day about a month later, I was having lunch with Borge at a local restaurant in Renton. We chatted endlessly about how this HGS option would work for the BBJ. During our meal, I explained to him what had happened to Mike Hewett, and how I was worried that he would not be an HGS fan. Borge did not seem worried in the least. Instead, he took a napkin and wrote something on it. Then he folded it up, reached over, and put it into my shirt pocket. As I reached for the paper, he gently slapped my hand and told me not to look at it until I got back to Oregon.

After lunch I got into my car. I just couldn't wait any longer. I retrieved the napkin, unfolded it, and glanced at what was written. It was the price

Borge was prepared to pay for each HGS to make it a BBJ customer option.

It was similar to what Alaska had paid for a larger quantity, but I thought John and Jim would be able to accept it, given that it was a pathway into Boeing. I was able to call Borge the next day to let him know that our management had agreed to his price, and to give him our thumbs up on the project. During the call, Borge told me that if I could get three-quarters of his initial customers signed up for the HGS, he would make it standard equipment on his new aircraft.

That was a BIG incentive!

Our A-Team now had a firm commitment for the HGS on the new B737NG and decided a new name was needed. The B727 had been the HGS-1000 and the Dash 8 was the HGS-2000. They decided on HGS-2350.

Something else was happening almost simultaneously that would change things not only at Boeing, but in the aviation industry!

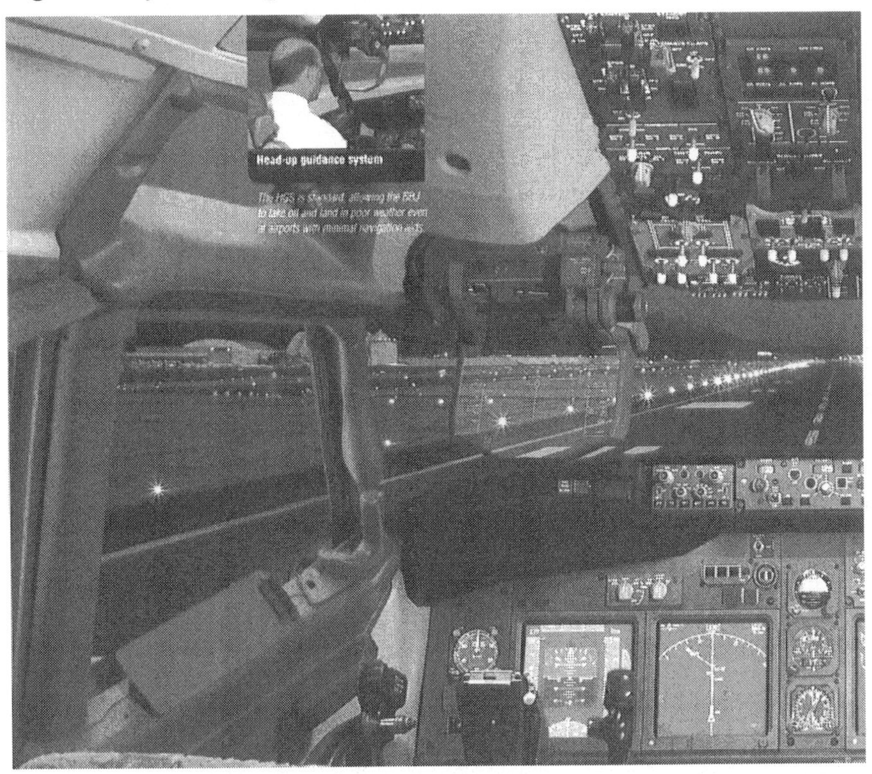

The HGS-2350 in the new Boeing Business Jet as an option
Inset, Mike Hewett flying the sim post-injury

The HGS-2350 installed in the newest B737NG "glass cockpit"

The B737-300, or Classic, with a "round dial" cockpit

Chapter 11

A BIG Customer

We were getting a lot of good press over a July 1991 Flight Safety Foundation study that had researched aircraft accidents over the period from 1959 to 1989. More than a thousand accidents had been analyzed to determine if our HGS could have helped avoid any of the events. After extensive investigation into the causes and NTSB findings, the study concluded that an HGS would have prevented, or "positively affected the outcome," of more than a third of the accidents. This was amazing, and really bolstered our safety argument in favor of the HGS. Many of our Boeing team supporters were now touting this report internally as yet another benefit of the system. We ordered reprints and they became an integral part of our sales and marketing arsenal as we chased after other OEMs and airline customers.

While all the activities with BBJ and Boeing were keeping me quite busy, Captain Sunshine was not idle with respect to other possible airline customers. I was keeping an open mind. One day in early 1993, while at my desk in Oregon thinking great thoughts, my phone rang. Our receptionist announced that she was putting through an interesting call for me.

A friendly voice introduced himself as Troy Menken, an engineer with Morris Air, an airline based in Salt Lake City, Utah. According to Troy, Morris operated eighteen B737-300s. His boss had asked him to contact our company to get some information about the HGS, including the rough price. After I'd answered his questions, he asked me to come to Salt Lake City to give a technical presentation on the HGS system and benefits.

This was for a Boeing aircraft still in production!

I arrived in Salt Lake City the night before our meeting, and Troy and I got together for dinner. We hit it right off and he provided me the lowdown on Morris Air. It was started by June Morris, wife of *the* Morris of the Morris Salt Company. She had established her own holiday travel business, but was sick and tired of the hassles with regular airlines and wanted a dedicated carrier for her customers. So she started Morris Air and hired a fellow by the name of David Neeleman to run the airline for her. It was David, president of the airline, who had asked Troy to call Flight Dynamics.

The next morning, after I arrived at the Morris facility, Troy took me down the hall and introduced me to Randy Redfern, their Chief Pilot, who was more than a little interested in the HGS, Skip Pennyweight, their Director of Training and Standards, and others. They had all been reading recent articles in the trade magazines that were very complimentary to our system. They had also seen the Flight Safety study that I had mailed to Troy after our initial call, and it seemed to have a big impact on their thinking. I met Usto Schultz, their avionics specialist, who would be heavily involved in any retrofit if they decided to proceed, and I was also introduced to David Neeleman.

The presentation to the group assembled by Troy, which included Neeleman, went very well. I was getting better at it with each successive customer meeting.

Several interesting things popped out during this get-together. The first was that the Morris B737s were already equipped with the IRS and DADCs, so that cost was not an issue. This was great news.

But the aircraft also came equipped with a Boeing-installed CATIIIa autoland system, making a large part of the HGS capability redundant. When I asked the group why they were so interested in our HGS, they explained that they had not been maintaining the aircraft's autoland to the required CATIIIa certification standards. The cost of bringing the system up to snuff was well over a hundred thousand dollars per aircraft—substantial enough for them to consider the HGS alternative. They also felt confident that our HGS was likely to be far less maintenance-intensive than the autoland. Another light went on! I hadn't thought about maintenance costs as a selling point, and I didn't know that autoland maintenance to CATIIIa was so expensive.

On top of all this, they were now flying more often to the Pacific Northwest and the West Coast of the US, where fog was frequent.

David seemed to fully understand the system's many and sometimes obscure benefits, and he appeared almost ready to sign up for the HGS at our first meeting.

While these events were taking place, Alaska Airlines, with the help of Dick and the A-Team, had achieved certification of lower landing minima at a CATI facility with the HGS—from 2400 RVR down to 1200 RVR. They were also submitting our application to the FAA for 300 RVR takeoff with the HGS, a reduction from the current 600 RVR. These were benefits that Morris Air was quite interested in.

The FAA seemed to fully understand that airline customers had to justify the purchase of our system, and they firmly believed in its safety advantages. "It will also keep aircraft out of smoking holes," one of our FAA supporters told me (after telling me I could not use that quote).

Compared to our previous HGS efforts, this campaign went lightning fast. Shortly after a single M-Cab session for Randy and Skip, and within five or six weeks of my original visit to Salt Lake City, we had a contract for eighteen systems with some additional spares. I couldn't wait to tell the Boeing team. They were surprised that a customer that already had the

CATIIIa autoland would choose to also take the HGS. I decided not to bring up the expense of upgrading the autoland system as one of the primary reasons.

The Morris Air win would go a long way to help change the mindset at Boeing.

Morris Air decided they could do the aircraft HGS installations at their own facility in Salt Lake City. After we'd achieved our first B737 HGS-2350 STC, members of our A-Team helped with the actual HGS installations.

They had completed about eleven of the eighteen aircraft in the Morris Air fleet of B737s when I got an alarming telephone call, from David himself.

"Phil, we are going to have to halt HGS installations until a big issue here is resolved," he told me.

My heart sank.

"Is it an issue with our HGS performance?" I asked.

"No," he responded. "I will call you within a week to explain the details."

This could *not* be happening again! It reminded me of the dreadful calls I had received from Larry Hecker at Western, and later from Dave Haapala at Northwest.

Even Captain Sunshine was despondent.

Jim and I once again went through a bunch of the possible scenarios that were becoming all too familiar. All of them were ugly.

Finally, David called me and gave me the news.

"OK, Phil, here's the scoop. We either have to remove eleven HGS from our fleet...or we have to add two hundred and thirty six more."

What?!

He laughed heartily at my reaction, and then went on to explain that Southwest Airlines, based in Dallas, Texas, had made an offer to acquire Morris Air and it had been accepted by their board.

At our sales meeting the next week, John advised me that I was going to "live in Dallas" until we closed the deal with Southwest.

This was to become a major game-changer at Flight Dynamics and Boeing and, indeed, for the whole commercial airline world.

As part of the Morris acquisition deal, David was now on the Southwest board. He offered to introduce me to some of the key players at the airline there in Texas. I immediately made plans to fly to Dallas, where I met John Owen, CFO, and Paul Sterbenz, VP Flight Ops. I especially liked Paul. I also met Gene Stewart and Prewitt Reaves in engineering, and Joe Marott, Director of Training at Southwest. Joe and I saw eye-to-eye on just about everything, and we also became good friends.

I ended up doing multiple presentations to flight ops, standards, training, engineering, and other groups at the Southwest facilities in Dallas over the next few weeks. They asked a wide variety of questions, which sometimes required answers and support from our A-Team. The things I learned from those questions and answers were to help me with future airline campaigns.

If the airline decided to go forward with the HGS, Marott would face the challenge of training hundreds of their pilots—a daunting task—and installing the HGS in a large number of flight simulators at their facility.

Meanwhile, there was no need for M-Cab simulator sessions in Seattle because Southwest technical pilots and Flight Ops management could fly the HGS on any of the HGS-equipped Morris Air B737s they had acquired (they had approval from the FAA to do this). Many of them became fanatical HGS supporters after flying the system.

Interestingly, Southwest had the same problem as Morris Air—they had not maintained their autoland systems to CATIIIa status on their fleet of B737-300s and -500s. They hadn't needed the systems, and didn't want to pay the maintenance costs, so they had actually been disabling the systems and removing the aircraft auto-throttles, which were an absolute necessity for the autoland to function properly, and giving them back to Boeing.

With the acquisition of the Morris Air network, they were now expanding their route system to the West Coast of the US, including the Pacific Northwest, both areas with lots of fog. Their low-cost airline philosophy included very tight turn times at the gate, so any weather-related delays or disruptions played havoc with their schedule. The cost of upgrading the autoland and re-installing the auto-throttles was substantial, and would require more aircraft down-time than HGS installation. I began to feel that this one was going to be ours to win.

On one occasion, John Owen decided to visit Morris Air headquarters in Salt Lake City, and spent a whole day with Troy Menken asking all kinds of questions about our system. Their meeting continued over dinner and went late into the evening, as Troy brought John up to speed on all the intricacies of the HGS. John became a true fan, which was quite unusual for a CFO.

David Neeleman later left the Southwest board and went to Calgary, Canada, where he started low-cost carrier Westjet. He then relocated to New York and began JetBlue, another airline success story. Finally, he moved to Brazil, where he had been born, and founded still another successful airline, Azul. He remained a strong HGS advocate and all of his airlines would eventually decide to go with our system. In fact, JetBlue later became the first airline to equip new aircraft with dual HGS from the factory.

There were many trips to Dallas in the months that followed. It was the first time we got to try out the new payback model we'd developed with the help of Tom Geiger and Dave Haapala at Northwest, and that had been uniquely refined by our own George Kanellis. Southwest was impressed. We were getting this payback thing down to a real science!

In the summer of 1994, while I was extremely busy with the Southwest campaign, Collins, part of the Rockwell Group, made an offer to buy fifty percent of Flight Dynamics from Hughes, who were now focusing on military programs. Kaiser Aerospace, a military HUD company that had partnered with Collins on another aerospace project, acquired the other fifty percent. Flight Dynamics became a Collins-Kaiser Company.

A contentious issue arose almost immediately for our marketing group. Collins was strongly opposed to any commissions being paid on HGS sales, since their own large sales and marketing team was not receiving commissions. Paying us commissions, they said, would create tension in the ranks of their own group. Needless to say, the HGS team strongly objected.

Collins and Kaiser gave notice that any deals that could not be closed in the next thirty days would not be eligible for commission. Thirty days? They had to be kidding! We were quite upset. I was risking the biggest HGS sale in the history of our company, and Dick had been working feverishly on Dassault, an OEM in France that was producing the F900 and F2000 business jets and was very close to a deal with them.

The thirty days disappeared in a flash, and with them, our hopes of any commissions.

A couple of months later, executives from both Collins and Kaiser visited our facility in Oregon to explain their acquisition to all our employees and to answer any questions we might have. We had set up a small stage in our production area, with a PA system and microphone. I was in my office getting ready to attend the speeches, when my telephone rang. It was Paul Sterbenz from Southwest.

I explained to him that I was running to a company presentation by our new owners.

"Well, Phil," he said. "I have some very good news for you. Our board has decided to proceed with the HGS project for the whole fleet. We are done—you got it!"

I wobbled a bit, feeling like I might faint, and sat back down.

"Is…is this for sure, Paul?" I stammered, barely able to contain my excitement. "I would love to be able to tell the new owners, as well as our own management, right now." I needed to be sure I had heard correctly. Paul confirmed that I was not deaf, or hearing things—I still had all my marbles.

I practically killed myself running downstairs for the presentation. John was getting ready to go to the mic. I told him I needed to make a short announcement. He became concerned and asked what about, but I said he would find out, and he would like it. I grabbed the mic, and announced that I had just heard from Southwest that we had won the HGS program there—a project to deliver over two hundred and thirty six HGS plus spares.

The room went nuts! There was cheering, hooting, and clapping like I had never heard before at our facility. John was incredulous.

What a perfect time to get this news. Our very happy new owners came over to congratulate me, and I reminded them what a team effort it had been.

We began the major work of spooling up for such a large order, from such a significant customer. Southwest was requesting things we had never been asked for before. They wanted an illustrated HGS parts catalog, for example, and much more. We got busy. This was a good one, and I knew immediately that it would have a huge influence on Boeing. I clearly remembered Alan Mulally's suggestion about bringing an airline customer with a significant order. Surely this was it!

While I did not receive my full commission, Collins and Kaiser agreed to a bonus, although it was a lot less than my commission would have been.

The next spring, some members of our management and I were invited by Southwest to join their summer LUV Classic in Dallas, an annual golf game and barbecue event. It was the first time I would meet Bob George, the Boeing customer engineer covering Southwest. I designed a T-shirt that intertwined the shape of the HGS combiner with a heart, the logo of Southwest (they were based at Love Field in Dallas). The caption read "LUV and HUGS" (for Head-Up Guidance System). Everyone loved the shirts!

We also scheduled a celebration event with the key players from Southwest for the next evening. We called it "the Steakout." I had a cake made with the "flat-footed duck" on it.

As the Steakout wound down, Paul came over and gave me more good news, as if that was even possible. Southwest had about a hundred and seventy B737-300s and some newer B737NGs (Next Generation) on order and scheduled for delivery from the Boeing factory. He told me that Southwest had no intention of retrofitting over two hundred brand new aircraft with our system after they were delivered from Boeing. They were planning to request—no, demand—that Boeing install the HGS on these new aircraft right at their facility in Seattle, prior to delivery.

This was truly unbelievable! I told Paul he had made my day…no, my week…hell, my whole year!

Meanwhile, Southwest still had all those existing aircraft to retrofit. While living in Everett, Washington, I had visited an aircraft modification center called Tramco, at nearby Paine Field airport. Alaska had earlier worked a deal with Tramco to do the HGS installations on their entire B727 fleet. I had met Al Dirvanowski, one of Tramco's senior engineers, while I was checking on the progress of the Alaska installations. I visited him again and told him about the Southwest order and the quantity of aircraft we needed to retrofit. He almost fainted from excitement.

I told him that we needed a rough cost of the per airplane HGS installation for the existing Southwest B737 fleet. He said that Tramco still had a lot of notes on the installation requirements for the Alaska aircraft. The cockpit structure of the B737 was identical to the B727, which meant that the HGS attachment hard points were also the same, but he requested drawings that showed the wiring and other differences from the B727, and told me he would ask Southwest for permission to visit and inspect one of their aircraft.

About two weeks later, Al called and gave me his estimate—it was around thirty-five thousand dollars per aircraft. I passed the number on to Paul Sterbenz and Gene Stewart in engineering at Southwest. That's where Bob George, whom I had met at the LUV Classic, came into the picture.

Gene had asked Bob to find out what Boeing would charge to do the factory install of the HGS on their new aircraft prior to delivery. Bob checked around and, after quite a few meetings at Boeing, came up with a price that was over ten times the Tramco estimate, despite the fact that installing in a new aircraft while it was being assembled was much easier than retrofitting. Gene flipped at the price, as did we.

Bob, with a lot of help from Jim Von Der Linn, worked magic with the Boeing team in an effort to get close to the price quoted by Tramco. This did not sit well with some at Boeing, but more on that in the next chapter.

Southwest had announced that it was the launch customer for Boeing's newest B737NG, with a firm commitment for a hundred of the -700 model aircraft. Their intent was to eventually replace their entire B737-300/500 fleet with the new variant. Paul told me that the -700s were part of their backlog at Boeing, and Southwest had added them to the list of required HGS installations.

The news just kept getting better!

Jim Von Der Linn took on the task of getting the HGS offered as a Boeing factory option at a reasonable price. There was significant pushback by some Boeing engineering management, who felt that Boeing shouldn't install anything in their airplanes that they hadn't designed or specified and engineered.

To help reduce the HGS install cost, Flight Dynamics had offered to give Boeing, at no charge, the FAA-approved HGS STC that we had achieved for the Morris B737-300 aircraft. For the new NG aircraft, we had the commitment to do Borge's BBJ, so we knew there would soon be an STC for the NG version. We promised to provide that to Boeing as well when it was completed.

Jim Von Der Linn and Bob George persisted, and were eventually successful in lowering the install cost. In calculating the factory install price, Boeing wanted to factor in the cost of the aircraft downtime that would be avoided by doing the installation before delivery. The final Boeing factory HGS installation price for each B737-300/500 was just under a hundred thousand dollars, a number acceptable to Southwest.

We would eventually also have to modify the HGS-2350 symbology to emulate the Honeywell head-down glass cockpit displays on the new B737NG. Boeing insisted, as Bombardier had previously, that for safety reasons the head-up and head-down symbology had to have a consistent look. We fully agreed.

In August of 1995, Boeing delivered the first factory-HGS-equipped B737 to Southwest!

The suggestion from my Boeing comrades and their senior management that we bring a significant customer to the table had done the trick. Southwest lived up to their promise to get Boeing to agree to installation at the factory and Boeing, with the significant number of aircraft in question, agreed to offer our HGS wiring and structural provisions as a factory option. This allowed rapid post-delivery installation of the HGS itself, and would eventually garner many more HGS airline customers for us.

Boeing's decision was a major game-changer.

Southwest was well known in the aviation world as a frugal, low-cost carrier, so their commitment to our HGS carried even more weight than we originally anticipated.

About this time, I learned that Gulfstream in Savannah, Georgia, was becoming more interested in HUD. They had a family of large business jets well known the world over as the cream of the crop. John Desmond had provided me with a couple of names at their facility. I called, and was invited to Savannah to give my HGS-101 pitch to the team there.

John was especially interested in this one and accompanied me on a couple of later visits to their facilities. Pres Henne, their VP of Engineering, was key to getting approval for the HUD project there.

GEC Marconi, a military HUD company based in the UK, had been monitoring our growing success in the commercial HUD world. They had partnered with Gulfstream's cockpit supplier, Honeywell, and were our primary competition for the project. GEC planned to supply the overhead projector unit and the combiner, and Honeywell would develop the required HUD computer.

At Gulfstream's request, I had worked up a proposal for the model GV bizjet that included our per system cost and our Non-Recurring Engineering (NRE), which covered our development and final FAA certification costs for the system on their aircraft.

One day, while I was back at our plant, Pres Henne called me.

"Phil, we are very close to a final decision on the HUD supplier for our aircraft," was his news.

"That's great," I responded. "So what's next? Do you need any more information from us?”

“Well, your competitor has managed to reduce their NRE to zero, and if Flight Dynamics can do the same, you will win the program,” he replied.

It just so happened that at that moment, Desmond and Caliendo were in our main conference room with representatives from both of our new owner companies for a board meeting. I asked Pres if he could hold on for a few minutes. While it was never a good idea to interrupt a board meeting, I barged into the conference room. Quickly, I explained the situation and that if we could “eat” the NRE on this project, it was ours. The answer was “No.” This was quite a surprise to me, since I knew that Desmond was very interested in this OEM. I tried hard to convince the group of the overall sales potential, but no deal. I was devastated, and returned to my caller from Gulfstream with the news.

“Too bad, Phil,” Pres said, “this could have been a good one for you.”

The competitor was selected. I strongly felt that we had made a terrible blunder. We were not to hear the end of GEC Marconi.

Shortly afterward, I got a call from EJ Carleton, VP of Safety for UPS in Louisville, Kentucky. While I had been busy with the Morris Air and Southwest campaigns, Jim Gooden had spent a lot of time with the UPS technical and flight ops teams in Louisville. He had presented the HGS-101 for their good-sized fleet of B727s, so the groundwork was already complete. Jim was busy with developing an overall sales and marketing plan for our new owners, and asked me to take over at UPS.

In my first meeting with the UPS team, I answered their questions and gave them references to Tom Johnson, Dean Schwab, and others at Alaska Airlines. We now had about thirty surplus HGS-1000s that Dick had managed to buy back from the incomplete F27 project, and UPS became quite interested when I said we could offer them a steep discount on enough units for half their fleet. After some dickering over the price for the remainder of the B727s, they signed up for sixty systems for their B727 aircraft freighter fleet, which included the thirty surplus units. Another win, thanks in great part to Jim’s early efforts.

After some office celebrations and beers at our favorite pub, I continued my Boeing campaign.

But there was more ugly on the horizon.

FLIGHT SAFETY FOUNDATION
SEPTEMBER 1991

FLIGHT SAFETY
D I G E S T

Head-up Guidance System Technology (HGST) — A Powerful Tool for Accident Prevention

Project Report
FSF/SP-91/01
July 1991

Special Safety Report

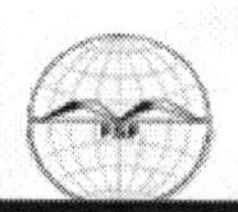

The Flight Safety Foundation study on HGS and accident prevention

The A-Team and the Morris Air B737-300 HGS cert team

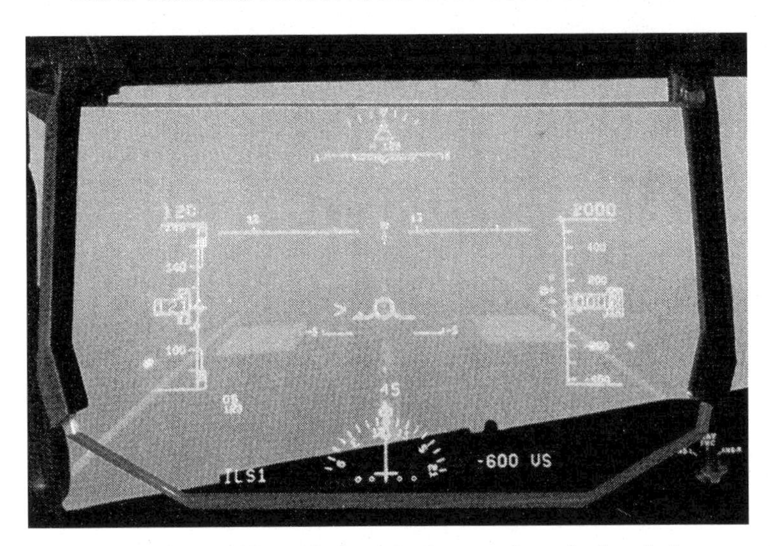

The B737NG HGS symbology had to emulate the head-down primary flight display (PFD) symbology

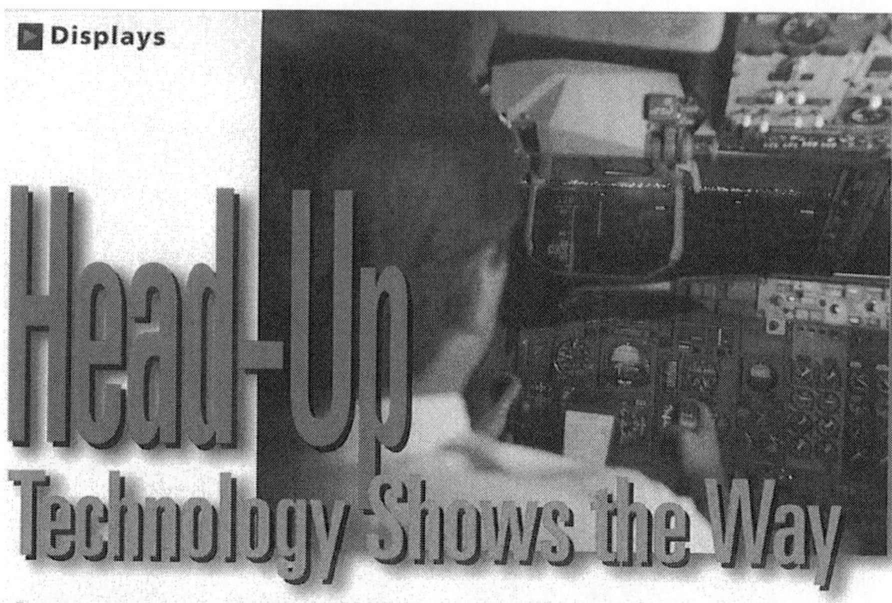

Southwest Airlines is reaping both economic and operational benefits from the Flight Dynamics Head-Up Guidance System. by Charlotte Adams

HEAD-UP GUIDANCE Systems (HGS) have been making a splash in civil aviation ever since Alaska Airlines pioneered the devices on its Boeing 727-200s. Alaska and HGS manufacturer Flight Dynamics, the Collins-Kaiser venture based in Portland, Ore., cleared the technology for Cat. IIIa weather minimums, setting an industry precedent.

From relatively modest initial sales, HGS systems are becoming a familiar sight. Southwest Airlines, in the largest head-up technology buy in commercial aviation—a $45-million contract announced in August 1994—will equip its B737-300s, -500s, and prospective -700s with new displays.

It's been a year since the airline's first systems were approved. From a base of eight head-up systems acquired through its purchase of Morris Air, Southwest plans to deploy 274 of the devices by 2001—and more if options for additional aircraft are exercised.

The Flight Dynamics HGS system, built around a wide-field-of-view, holographic head-up display (HUD), was designed to allow for Cat III operations with landings in visibility as low as 700 feet and takeoffs in visibility as low as 300 feet. It consists of six line replaceable units (LRUs): an HGS computer that receives signals from aircraft sensors and converts the data to symbology; a control panel for setting glideslope angle, runway length, and elevation, and for selecting modes of operation; a drive electronics unit that conditions signals from the HGS computer to drive a cathode ray tube in the overhead projector; the overhead unit, located above the left pilot's seat; and the combiner, which reflects the light projected from the overhead unit. An annunciator panel provides system status and warning indications.

In addition to signal processing, the HGS computer evaluates system and approach performance through Built-In Test (BIT), input validation, and approach monitor processing.

Advance Billing

According to officials at Southwest, the performance of the system is living up to its advance billing. Last December, for example, HGS-equipped Southwest 737s were landing at Sacramento, CA—one of the worst U.S. airports for visibility—with all flights on time, said Beth Harbin, a company spokesman. Moreover, she added, "We were accommodating people from other airlines."

In total, at all airports the airline operates to, in 11 months last year, 7,292 approaches were flown with no go-arounds, although the majority took place in Cat. I visibility.

To achieve these types of results so quickly required commitment from other parts of the organization as well. "One hundred percent of the pilots are now trained," reported Capt. Milt Painter, Southwest's director of flight standards and training.

Avoiding flight delays and diversions is especially important for Southwest because of the way it flows airplanes, said Joe Marott, director of Southwest's

Courtesy Avionics magazine, March 1997

For Immediate Release

Contact: Karen Swanson, Flight Dynamics
(503) 684-5384

AUGUST 25, 1994 (Portland, Oregon): Flight Dynamics Wins Major Order From Southwest Airlines.

Flight Dynamics today announced that it has signed a contract with Southwest Airlines Co. to supply Head-Up Guidance Systems (HGS®) for Southwest's fleet of Boeing 737-300s, -500s and -700s. The contract, valued at over $45 million, includes firm orders for 236 systems and additional options. Flight Dynamics, located in Portland, Oregon, is jointly owned by Collins Air Transport Division of Rockwell International, and Kaiser Aerospace and Electronics Corporation.

The HGS order by Southwest will permit the airline to conduct Cat III operations with landings in visibility as low as 700 feet and takeoffs in visibility as low as 300 feet. The HGS also increases operating capability at Cat I airports.

John Desmond, President of Flight Dynamics, said "We worked with Southwest to conduct a comprehensive analysis that determined passenger benefits and the economic advantages of employing Head-Up Guidance on the airline's routes. Being able to land and takeoff in low visibility will reduce the number of cancellations, diversions, and delays due to weather. Southwest was convinced that the financial return to the airline and the benefits to their customers combined to justify the investment."

Including the Southwest order, Flight Dynamics currently has orders for its HGS valued at more than $100 million. In addition to supplying the HGS for the Boeing 737 series aircraft, Flight Dynamics has delivered systems for the Boeing 727. The company has orders for and is conducting HGS certification programs for the Canadair Regional Jet, the deHavilland Dash 8, the Dornier 328, the Dassault Falcon 2000 and the Saab 2000. Lockheed has also selected the HGS as standard equipment for the new Lockheed C-130J.

###

P.O. Box 230609
Portland, Oregon 97281-0609

16600 S.W. 72nd Avenue
Portland, Oregon 97224-7799

Phone: (503) 684-5384 FAX: (503) 684-0169

Chris, Joe Marott, and me (in the Luv & Hugs t-shirt)

Bob George and me celebrating at SWA's LUV Classic

John Desmond and Paul Sterbenz at SWA's LUV Classic

Southwest asked for our sticker art and came up with their own.

The Southwest Airlines-Flight Dynamics "Steakout" team (l to r): Joe Marott, John Owen, John Desmond, Paul Sterbenz, and me

The "flat-footed duck" cake

A Whole New Approach

As most of you know, our Southwest aircraft are in the process of being equipped with a device called the Head-Up Guidance System (HUD) to enable our Pilots to land where there is as little as 700 feet of runway visibility which is known as Category III. Before the installation of this equipment we were classified as Category I requiring 1,800 feet of visibility to land. The installation of HUD means fewer flight delays, cancellations, and diversions due to weather. And, of course, the most pronounced use of HUD is in the Pacific Northwest where dense fog is an operational challenge.

Installation and training began in March 1995, and by December of this year, the job will be complete. The Head-Up Guidance System is a great asset for Southwest Airlines, and after reading Captain Brian Anton's personal account of his first Cat III landing utilizing the HUD system, we think you will readily agree. And a special thanks to Brian for sharing this experience with us.

We arrived in Burbank at 7:35 a.m. local and were advised to call Dispatch who wanted to know if we were Cat III qualified. I checked the weather sheet, and, sure enough, the weather on our outbound flight to Sacramento was one-sixteenth of a mile. Our aircraft had just been placarded Cat III two days prior. Enroute to Sacramento, I briefed the Flight Attendants about the Head-Up Guidance System and the possibility of an aggressive go-around and even the wheels touching the runway if initiated at the lowest minimums. They appreciated the warning, but I said this was unlikely because the weather was improving. I also took the opportunity to brief our Customers and brag about our new HUD system.

Things changed as we got closer to Sacramento. The radio was filled with reroutes, holding instructions, and diversions. Nobody was getting into Sacramento. We itched to Approach and heard the bad news, Southwest 896 Sacramento RVR 800-800-800-, say intentions". I told him we were Cat III and could go to 700 RVR. Boy, did that feel great! As we descended on downwind, the RVR went to 700. This meant a real live Cat III approach hand-flown down to 50 feet above the runway.

I started hand flying the HUD on downwind so I could get the feel early. Jim's (First Officer Jim Robertson) procedures were outstanding and made it very easy for me. This is a real tribute to our Training Department for a First Officer who hadn't seen a HUD landing since his simulator training.

When we turned onto final at 3,000 feet, the sun over the fog bank was almost blinding. I configured to flaps 40 about three miles early so I could trim the airplane perfectly. I was surprised that even in the bright sunlight I didn't have any trouble seeing the display.

At 150 feet above the runway, it still looked like white butcher paper across our windshields. Jim stated, "Approaching minimums". Then, just a few seconds later a most amazing thing happened. At the moment we hit minimums (50 feet above the concrete), Jim said, "minimums." Simultaneously, I saw the runway centerline lights and said, "Landing." The display I had seen was incredible. The bars simulating the sides of the runway on the HUD display were lined up perfectly with the actual runway: and right in the middle of that foggy display, the actual runway centerline lights came into view right where they were supposed to. We touched down at the 1,000 foot mark-dead center on the runway.

I could hardly see to taxi, the fog was so thick. Tower and ground control may as well have been working from a phone in their living rooms-they couldn't see a thing. We crept our way to the gate with a smile I haven't been able to get rid of for days. We were almost at the gate when the surprised Ramp Agents scurried out to meet us. One told me, "We didn't know you were here-we didn't see a thing."

I went into the jetway to tell Dispatch of our success and to come up with a code for a "HUD appreciation delay." Our Customers were lined up down the aisle like Disneyland, each wanting to get into the cockpit and take a peek at the HUD. I wish I could express the wonderful reaction of our Customers, They were so impressed!

Then, things got even better. Other airlines lost about 70 passengers to us since their planes could not get in. As we rounded up these *very* appreciative passengers, I noticed a television crew taking shots of us through the terminal window. Of course, they didn't have much else to look at but our red-bellied bomber and a few hundred feet of fog. I asked the Agent to invite them to our cockpit to see our new HUD landing system. Boy did they jump at the chance! The female reporter was so impressed that she giggled like a little kid when she leaned over and saw the actual HUD display. I did the best I could talking about our new HUD and our wonderful airline. In the back of my mind, however, all I could think of was, "God, I wish I had gone on a diet and gotten a hair cut!"

I just can't fully express how impressed I am with the HUD. Even the reaction of our Customers was way beyond what I expected. Whatever thoughts I might have had about hand flying an approach to 50 feet above the concrete have turned into a desire to fly every one I can.

I know our Senior Management team are the most frugal people in the airline industry, and that's essential to our success. I just want to say THANK YOU for spending money on the HUD system. I think the pluses will be greater than anyone ever imagined. We flew five legs that day and finished ontime. That's just unbelievable for such bad weather. From the Customer Service Agent that was able to board 137 very happy Customers out of a terminal filled with frustrated, delayed people to the p.m. flight crew who got our airplane ontime-the HUD system was appreciated and will definitely assert our position as "the best deal in town."

March 1996 - Luv lines
A Corporate Newsletter for Southwest Airlines Employees

An amazing article published in the Southwest Employee Newsletter

Collins Avionics Report

MARCH 1996

Flight Dynamics Head-Up Guidance System Chosen for UPS 727s

United Parcel Service has ordered 60 Flight Dynamics Head-Up Guidance Systems (HGS®) for its Boeing 727 aircraft. The HGS will permit UPS to land its aircraft in visibility as low as 700 feet and to take off in visibility as low as 300 feet, significantly enhancing its aircraft operations in low-visibility conditions. Flight Dynamics is jointly owned by the Collins Air Transport Division and Kaiser Aerospace.

Use of the HGS results in fewer flight delays, diversions and cancellations due to low visibility. According to UPS Airlines President and Chief Operations Officer Tom Wiedemeyer, that allows UPS to be even more reliable. "With HGS, our aircraft will be in the air taking packages to their destinations, not waiting on the ground for weather to clear. Because we are using the most advanced technology, UPS customers can be assured that UPS is doing everything possible to ensure that packages get where they need to go, when customers need them."

Flight Dynamics manufactures its holographic Head-Up Guidance System for air transport, regional airline, and corporate aircraft. While flight crews previously needed to divert their attention to panel-mounted flight instruments at critical approach altitudes, the HGS projects flight information, and a graphic representation of the runway, in the pilot's forward field of view. This information includes attitude, altitude, airspeed, flight path and the command guidance necessary for low-visibility operations. The HGS can also detect the presence of windshear and provide pilots with an emergency flight profile to guide them safely through the occurrence.

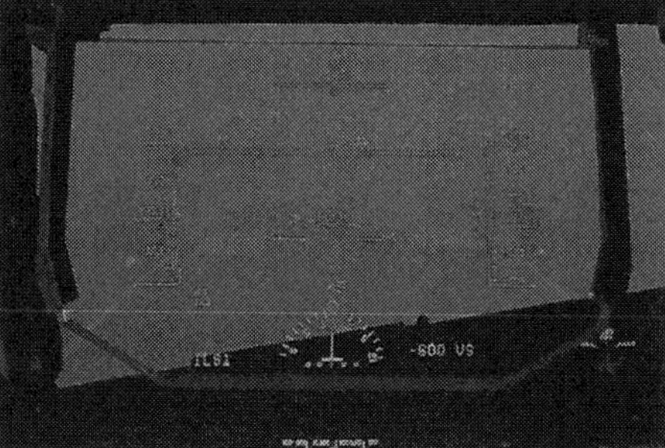

With its selection of the Flight Dynamics Head-Up Guidance System, UPS will be able to fly when poor visibility leaves other carriers stranded on the ground.

According to Flight Dynamics President John Desmond, "The UPS order confirms the universal appeal of our HGS, as it will now be employed in five major market segments, including package carriers, major airlines, regional airlines, corporate operators, and in military transport operations," he explained. "More people and more packages will experience on-time service in reduced-visibility conditions. This order pushes sales of the HGS to more than 700 systems."

The Flight Dynamics HGS first entered commercial operation on a Boeing 727 in 1989. Airlines currently taking advantage of the HGS technology

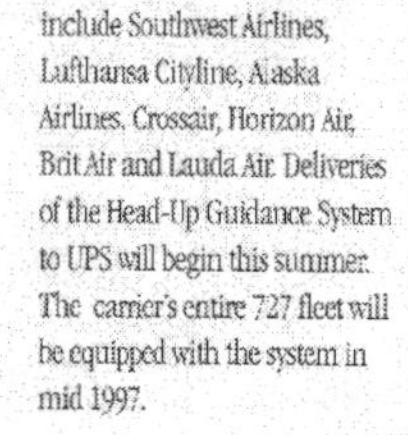

include Southwest Airlines, Lufthansa Cityline, Alaska Airlines, Crossair, Horizon Air, Brit Air and Lauda Air. Deliveries of the Head-Up Guidance System to UPS will begin this summer. The carrier's entire 727 fleet will be equipped with the system in mid 1997.

Chapter 12

Moving the Giant

Not long after the Southwest celebration parties had calmed down we heard that our good FAA buddy Berk Greene, one of the strongest advocates for the HGS, had passed away—he was only in his mid-fifties. I and many of the folks at our facility were devastated by the news. He was a true industry visionary, and without his herculean efforts in Washington, DC, to help us get the HGS certified, our story could have ended before it began. He fought an uphill battle all the way, with innovative and forward thinking, modifying many of the agency's rules and regulations to include the HGS for CATIIIa certification. He firmly believed in all the safety advantages of our system. His death was a terrible industry loss and a huge blow to all of us who knew and respected him.

We forged on with the task of changing aviation.

The commitment to our HGS from Southwest was an enormous boost, both for Flight Dynamics and for our new owners, mostly because of the impact at Boeing. Southwest was a large and influential Boeing customer, with an all-737 fleet. Here they were, insisting that Boeing install our system on a hundred and seventy new B737 aircraft coming from the factory in Seattle, as well as on the B737NGs they'd ordered.

We eased some of the internal engineering burden by offering to provide our Morris Air HGS STC at no charge to Boeing. Still, this project would require the OEM to go through some huge gyrations—there was no precedent for Southwest's request.

But there was another project inside Boeing that was to help us. A company called SBC (Scandinavian Bellyloading Company) had certified and installed a cargo handling system on some of Scandinavian Airlines' planes. It had been STC'd by the supplier, and other Boeing customers were now requesting it on new aircraft coming from the factory. The B737 and B757 badly needed cargo handling systems, and it made sense to find a way to offer airlines this solution.

With help from Sales, Jim Von Der Linn, who was in Renton engineering, had volunteered to work on a Boeing internal project called "Market Driven Target Cost," with the objective of ensuring that Boeing could meet cost expectations for an aftermarket system that an airline wanted installed at the factory. The HGS was one of these aftermarket systems, and the Market Driven Target Cost project was how Jim and Bob George had managed to get Boeing to quote Southwest an acceptable HGS install price.

Boeing had now assigned a HUD buyer in their ranks—this was yet another great sign for us. His name was Lyn Sorenson. I decided I had to meet him in person, so on my next trip to Renton, I arranged to meet Lyn for lunch, and I liked him very much. He was impressed that we had managed to sell our system to such a significant customer as Southwest, and assured me that Boeing would indeed have to fulfill Southwest's factory install request. He asked me many questions about the HGS and our company.

About a week after I met Lyn, his boss, a fellow by the name of Viggo, called to say that he and some folks from Boeing's procurement and quality assurance (QA) groups wanted to visit us to inspect our manufacturing facilities in Oregon.

Sure, no problem… Well, maybe a couple!

We hadn't yet ramped up our production for the Southwest order, but we had completed all our deliveries to Alaska for the B727, and to Horizon for the Dash 8. So the production line looked a bit skinny. OK—a lot skinny! I came up with a solution that was maybe a little questionable.

I asked some of the other folks in the plant, including a few secretaries, to show up at our soldering and assembly tables wearing the blue coats of our manufacturing staff. We placed them at the back of the area in case there were any difficult questions from our visitors. I also had our shipping department assemble a lot of empty HGS boxes and stack them on shelves near the delivery doors in clear view from the production area.

John, Dave Staehely (our operations director), and I met the visitors in the lobby and escorted them into our manufacturing area. Dave was in charge of answering most of their questions, especially about our supplier chain and quality control.

I had developed a number of explanations for the visitors in the event of the wrong person being asked any detailed questions. As expected, one of the Boeing QA folks walked over to a soldering table and asked the woman seated there what her job entailed. Luckily, she was one of the real ones, and was able to provide a very good, detailed answer. Whew! There were no questions about the boxes.

After the tour, we invited the Boeing folks up to the conference room for lunch. They seemed satisfied with their tour. I decided not to tell Lyn about our "Potemkin Village" until much later, at which point we both had a good laugh over a few beers. I was relieved!

A short while after the Boeing QA visit, Lyn called to let me know that he wanted to bring a group of Boeing engineers to visit our facility. They had some very specific technical questions for us. "The M-Cab simulator folks keep records, and our Boeing engineers have noticed that your HGS landings exhibit a very tight touchdown footprint," Lyn said. A touchdown footprint was a record of exactly where the actual aircraft wheels touched down on the runway.

"OK...," I responded, not yet understanding where we were going with this.

"Well, our popular B757 autoland is experiencing a few long landings," he continued. "Your touchdown printouts have totally amazed our Boeing engineering community, since your system is manually flown by the pilots."

"Yes," I replied. "So what do they need from us?"

He said the Boeing team wanted to ask some questions to better understand the HGS performance. He proposed a visit date.

The A-Team was instructed to attend the meeting. John knew that Doug would be key here, and asked him to please wear a suit that day. Doug hated suits and refused.

When the Boeing group arrived in the lobby I flew downstairs to greet Lyn and meet the rest of his team. I got them settled in our conference room with coffee and rolls. John, Dick, and the rest of the A-Team arrived...but no Doug.

After the introductions were complete, one of the Boeing folks explained that they were trying to understand the differences between the HGS touchdown footprint and the B757 automatic landing footprint. They were especially confused because our system required manual flying, so they expected to see a bigger touchdown dispersion for the HGS than for the autoland. Dick and Ken Zimmerman tried to explain, but eventually John leaned over and whispered to me to go fetch Doug. I went downstairs and found him in his office wearing jeans, a T-shirt, an old sweater full of holes, and sneakers. I knew John was not going to be happy.

Doug came up to the conference room and I introduced him to the group. John was scowling at him. Doug listened carefully as the Boeing folks explained their issue, but he said nothing. Finally, he got up, went to the whiteboard, picked up a black marker, and began to write.

It was a triple integration math formula. And it was long...really long.

He got to the far side of the board, the magic marker squeaking, came back, and began a second line. Then a third. There was silence in the room. Lyn and I were grinning at each other. We thought it was hilarious—our little company educating the Boeing Company.

The Boeing engineers were all writing down his lengthy formula as fast as they could. When Doug was finished, he put down the marker, went back to his seat, sat down, and said absolutely nothing. Finally, after a few minutes, he got up, looked at John, told him to call if he needed anything more, and left.

Unbelievable! I had to try very hard to stifle my chuckles.

Lyn never forgot that day. And we later had some really good belly laughs about it. Lyn told me that the Boeing team had talked all the way back to Renton about Doug's lengthy formula, and argued over what some of the symbols meant. But eventually, they seemed to come to agreement that what we were doing down there in Oregon was significant, and very high-quality work.

While this was going on, Borge had been providing me with a list of new BBJ customers. He warned me to be very careful, since many were Fortune 500 companies that did not want their purchase "up in lights." Borge was hoping that seventy-five percent of them would want the HGS equipment on the BBJ aircraft they ordered. I contacted them cautiously.

About eight months later, Borge invited me to the first-ever BBJ hospitality suite at the annual NBAA show in Orlando, Florida. This was the same show I had attended many years earlier at my first day on the job with Flight Dynamics. Borge approached me, shook my hand, and told me I had been successful—only two of the first twenty customers had opted not to take the HGS, and they were overseas, in countries where I had a hard time getting to them. He told me our HGS would now be standard equipment on his BBJ.

This was a first, and another in a long line of critical and significant events changing aviation.

Back to the "Big B" and factory installation.

Although some engineering managers were opposed to factory-installed aftermarket features, there was support for it at high levels in the company, and of course the sales guys wanted it. Alan Mulally liked and supported our HGS, but he did not like the idea of incorporating an outside STC into Boeing's factory build process. Some engineering managers, including Alan, believed that nobody else could design aircraft like Boeing could, and if you wanted to put something in an airplane, then Boeing had to engineer it. But the typical Boeing reengineering of outside STCs was extremely expensive and time-consuming, and the final numbers and timeline would typically be more than an airline customer could swallow.

Jim Von Der Linn, one of our most ardent HGS supporters at Boeing, was asked by his boss, Bob Hammer, the Director of Engineering in Renton, to find a way to fly under the opposition radar and get certified aftermarket products offered as factory options.

The company had lengthy shelves of manuals called Design Requirements and Objectives that detailed every aspect of designing and manufacturing Boeing aircraft. There was no process for getting something like our HGS installed in the factory. So Jim wrote a new process and managed to get it approved by every group in the company. Since the process affected all Boeing aircraft, he needed the blessing of both Renton (narrow body aircraft) and Everett (wide-body aircraft).

Boeing's Customer Engineers had the responsibility of working with airline customers to configure their new airplanes. These engineers helped customers select from the many options listed in the Boeing catalog. They couldn't just simply add an option to the catalog without what was called a "master change." Jim's effort significantly reduced the cost and complexity of developing master changes for systems like the HGS. Boeing no longer had to charge unreasonable prices to reengineer the installation of equipment that had already achieved an STC.

Now, for the first time, Boeing's airline customers could get what they wanted, such as our HGS, when they wanted it, and for a price they were willing to pay.

Some engineering managers were unhappy about what Jim had done, and he didn't get a raise for three years. But many people were ecstatic. Jim was invited to a Customer Engineering meeting where he was presented with a framed certificate signed by folks from Boeing, Southwest, and Flight Dynamics, thanking him for his "leadership, tenacity, and willingness to take the heat." (Bob George later received a thank you certificate as well.)

With the help of an amazing team, we had done it. The HGS provisions would now be included in the Boeing B737 customer options catalog. Absolutely incredible!

But there was another problem.

The Boeing B737 fuselage and the cockpit (referred to as "Section 41") were manufactured at their plant in Wichita, Kansas. The cockpit was mated to the fuselage there, and the completed assembly was shipped to Renton by rail. They were using the so-called "cookie cutter" method—all section 41s and fuselages were identical, so they just churned out the completed assemblies and let Renton decide which ones went to which customers later in the process.

The Southwest commitment was to force an unwelcome change at Boeing in Wichita!

The Southwest cockpits had to include the HGS installation hard points—the actual roof attachment points for the overhead projector and combiner. For the first time ever, the folks in Wichita had to label section 41s and fuselages that would be going to Southwest. This was a mess! It required some radical and difficult changes to the production process that many in Wichita did not understand or want to support.

In the interim, Alaska Airlines had placed an order for fifty B737NGs to replace their aging fleet of aircraft. After learning that Southwest was getting HGS factory installation at Boeing, Alaska also requested that their new aircraft be factory fitted with our HGS.

This event pushed Wichita over the top, since they now had two different customers whose section 41s had to be identified prior to shipment to Renton.

(They had some initial alignment problems as well, but more on that later.) I learned from reliable sources that Wichita had put out a "Wanted" poster with my picture on it.

Jim Von Der Linn got involved once again to save the day. He suggested to Boeing upper management that it might be less costly to just fit HGS hard point provisions to *all* section 41s, and Wichita could go back to their cookie cutter method. Turns out that was a very good and acceptable solution all around.

This turned out to be a hugely beneficial development for Flight Dynamics!

Now, Boeing's airline customers did not have to choose the HGS from the contracting get-go, because all the B737s coming off the line were essentially provisioned for HGS. It gave customers, and our team, more time to consider, and analyze, the benefits of adding our system to their new aircraft orders.

In the meantime, in my discussions with Lyn, I realized he was getting a little weary with his various positions at the Big B. I told him that maybe I could help.

Borge and I were meeting fairly regularly to assess our progress with new customers that might want Enhanced Vision Systems, or EVS, with our system (more about that later), and also to discuss some new HGS features that our company was developing. At my next BBJ lunch with Borge, I asked him if he might be interested in a Boeing fellow I thought would be a good fit for his new organization. I told him about Lyn, and he said he would check into it.

Around this time, we had decided to host annual HGS Operator Conferences for our customers. Because we had signed up both BBJ and Alaska, we scheduled our next conference in Seattle. I suggested to John that we arrange a boat tour with dinner on Lake Union, and he agreed. I asked our marketing assistant, Helene Bloch, to add Lyn's name to the invitation list. Lyn was surprised to receive the invitation, until I told him that I had mentioned his name to Borge for a position on the BBJ team.

I managed to get the two of them together on the boat tour, and Borge asked Lyn if he would accept his offer of a job in contracts with BBJ. Lyn jumped at the chance.

While I was preoccupied with Boeing and BBJ activities, the A-Team, had been busy...again! There were a lot of new developments in flight technologies, and they were working to get some new safety features into our HGS that would take advantage of these new developments. These included TCAS (the Traffic Collision Avoidance System), and also tailstrike protection (the B737-900, an NG variant, was quite long, and Boeing and the airline customers were worried about dragging the aircraft's tail on takeoff).

Dick had also been busy. Ever since my first trip to New Orleans to attend the NBAA for Flight Dynamics, I realized that there was significant interest from the business jet community—not just BBJ—and it could be a lucrative market for our product. Dick had taken on Dassault, a French business aircraft manufacturer that was hell-bent on becoming one of the early business jet OEMs to endorse and offer the HGS from their factory. They wanted the system as a customer option on two of their highly successful jet offerings, the F900 and the F2000.

Dick had worked the French end as well as their US subsidiaries in Teterboro, New Jersey, and Little Rock, Arkansas, and finally, Dassault became an HGS customer. Our marketing assistant, Helene Bloch, was a great assist to Dick and to our company, as she spoke fluent French. That turned out to be a blessing at the NBAA shows, where we would entertain Dassault pilots and senior management at our booth and, often, at dinners and receptions.

We were now regularly attending NBAA shows, and had developed a unique "half cockpit" simulator that was transportable. It was built by our A-Team at the Tualatin plant in Oregon, and was constructed from fiberglass to keep the weight down. It was a star attraction at the shows. We had the HGS installed with some preprogrammed routines that could be called up to allow visitors to see what the system really looked like in action. Often, there was a line of pilots waiting to try out our sim, and we generated lots of great leads and customer interest at the shows.

At one of these NBAA shows, a well-known actor approached us to have a look at our HGS. It was Kurt Russell, who had starred in a number of Hollywood movies. He was also a pilot, and had recently purchased a new Dassault F900, which now included our HGS as a customer option. He wanted to check it out before committing. We gave him HGS-101 and he seemed very impressed. While Kurt was looking into our system in the half-simulator, his wife, Goldie Hawn, was busy at the trade show picking out carpeting and leather for their new aircraft's interior.

Our new friend Russell opted for the HGS.

We were really having fun now!

Jean and Dick Hansen, Barry Raynor, Pete Freeland, George Kanellis, Jim and Kathy Von Der Linn

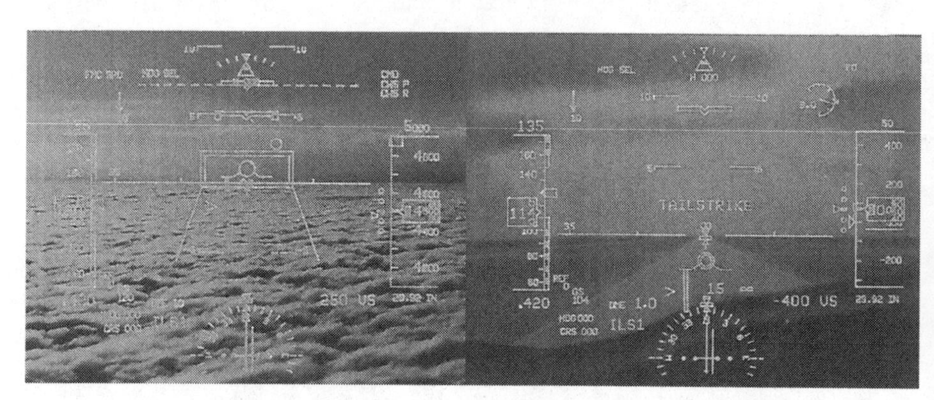

TCAS and tailstrike protection symbology on the HGS

Our HGS half-simulator with George Kanellis, me, Helene Bloch, and John Desmond

Tom Kilbane and me giving demos in the "HGS half simulator" at the NBAA show in 1995

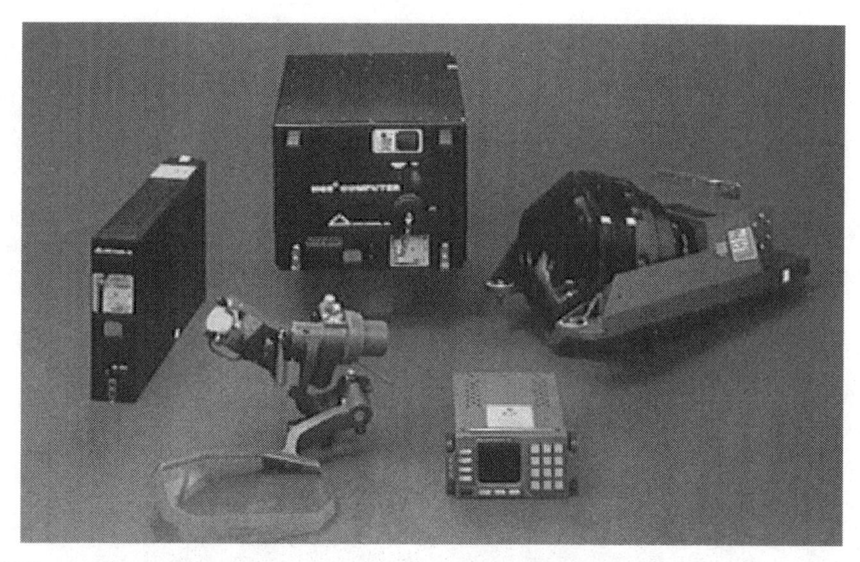

HGS components included the computer, drive electronic unit, combiner, control panel, and overhead unit

Next Step: Hybrid HUD/Autoland

The next step for lowering landing minimums below Cat. IIIa is to use two independent, fail-passive systems, exploiting both Head-up Guidance System (HGS) and autoland technology, experts say.

"This means that you will be able to have Cat. IIIb, or 300-foot runway visual range (RVR) landings," predicted Phil Moylan, Flight Dynamics marketing manager in Portland, Ore.

Flight Dynamics is working with Alaska Airlines to obtain FAA approval for such a hybrid system on its Boeing 737-400s, Moylan says. Alaska already has approval to take off at 300-foot RVR; the 300-foot landing minimum will give it a "balanced field approach," Moylan said. "Unbalanced" situations can "turn a hub into a parking lot."

FAA is updating its advisory circulars to address hybrid systems and hopes to complete the work by the end of the year. It is working with industry to develop "criteria that would allow those systems to be combined [and] potentially be authorized to 300-foot RVR" for landings, said Tom Imrich, FAA national resource specialist for air carrier operations in Seattle, Wash.

If and when hybrid technology receives the necessary approvals, it would "permit aircraft like the 737, which currently does not have a roll-out system installed, to be eligible for Cat. IIIb minimums," Imrich explained. "For new aircraft such as the Airbus A340, Boeing 747-400, and 747-500, sophisticated flight control systems, designed from the outset, will be able to do these functions with autoland."

He predicted that options such as HGS will be particularly attractive, in the near term, for retrofit (such as 747-100s and DC-10s) and commuter-class aircraft.

In the long term, HGS technology may be important for aircraft such as the High-Speed Civil Transport (HSCT), which may have no forward-view windows. The HSCT "may need to use multispectral capability to achieve the necessary accuracy, integrity, availability, and forward view for safe landings," Imrich said. —CA

Courtesy Avionics magazine, March 1997

A Dassault ad for our HGS customer option

Certificate of Achievement

Jim VonDerLinn

On behalf of Southwest Airlines, the Airplane Configuration Group is pleased to present this rare and deserved award.

In the spring of 1994, Southwest Airlines approached us to discuss the installation of a Heads Up Display Guidance System (HUD or HGS) for the 737-300 aircraft. This monumental task was destined for failure, due to a variety of reasons, with one exception. The exception was the will power and "can do" spirit demonstrated by you.

Despite entrenched cultural opposition to programs of this type, you persevered throughout the process to overcome all obstacles. Without your leadership, tenacity, and willingness to " take the heat", we would not have achieved the goal.

It is with great pleasure that we thank you for your performance and a job well done!

Thanks again for all your help Jim!

Many thanks

Jim
thanks for the great work
John D

Gary Scott
Vice President & General Manager
Renton Division

Luke J. Gill
Vice President Maintenance & Engineering
Southwest Airlines Company

Certificate of Achievement

Robert George

On behalf of Southwest Airlines, the Airplane Configuration Group
is pleased to present this rare and deserved award.

In the spring of 1994, Southwest Airlines approached us to discuss the installation of a Heads Up Display Guidance System (HUD or HGS) for the 737-300 aircraft. This monumental task was destined for failure, due to a variety of reasons, with one exception. The exception was the will power and "can do" spirit demonstrated by you, the Boeing employee.

After months of discussion and preparation, the first HUD aircraft was upon us. The task was greater than originally anticipated, but accomplished as one of the noteworthy challenges Boeing employees have conquered.

As the 737-300 discussions were taking place, Boeing was also conceiving the 737-700. As part of that conception, many considerations were being evaluated for HUD provisions. The necessity of those provisions was secured when Southwest began receiving the HGS on their -300 with the delivery of PS794, and then launched the -700 program.

Once again, Boeing employees stepped up to the challenge to satisfy the customer wishes. The amount of pro-active work directly with Flight Dynamics (the HGS supplier) and the supplier's willingness to participate here at Boeing has kept this program on course. The work with other related vendors has also been above average as noted by both outside suppliers and feedback from Southwest.

Accordingly, you should consider yourself as a World Class
and customer-oriented employee.

It is with great pleasure that we thank you for your performance and a job well done!

Gary Scott
Vice President & General Manager
Renton Division

Luke J. Gill
Vice President Maintenance & Engineering
Southwest Airlines Company

BOEING

Chapter 13

Bucket Parties

Over my many years at Flight Dynamics, I had always heard from Desmond that we needed one of the major US airlines as an HGS customer. I thought that would be Southwest, but John really meant United, Delta, Continental, or American.

We had hired another marketing type, Jim McKinney, to assist with our frozen turkey head efforts, and after extensive training on our system, he was given United.

In 1996, once Southwest was a done deal and it looked like Boeing was on a good track for factory installation of our system, I decided to take on Delta Air Lines. I figured it might be as slow a potential customer as Boeing had been.

I wondered if any of the ex-Western folks I'd met in LA were still around. I mentioned this to John, and he said he had a good contact at Delta who had shown interest in the past in our original HUD. His name was Harry Alger. He might know who to contact at Flight Ops Technical in Atlanta to start the investigation.

Harry told me I ought to chat with a Captain Carter Chapman at Delta. He was the MD-88 Program Manager, and was about to become Chief Technical Pilot for Delta. Harry gave me Carter's telephone number and I was off.

After I'd talked to Carter on the telephone for quite a while the next day, he suggested I come to Atlanta to discuss the HGS. Apparently Delta was looking into a large purchase of over three hundred of the new B737NGs! I told him that Boeing would now offer the HGS right from their factory on that aircraft. Carter said he would be on the B737 acquisition team at the airline and he was definitely very interested.

I had been to Atlanta, Georgia, a few times before on other aviation business. I'd found a unique little restaurant on Virginia Avenue, not far from the airport and Delta's headquarters, called Spondivits. I had visited their neat little thatched-roof establishment a few times and ordered their "Bucket of Shrimp Special" with garlic toast—an incredible meal with probably *the* best shrimp I had ever eaten anywhere.

After flying to Atlanta, I met Carter at the Delta Flight Ops building on their campus, gave him the HGS-101 presentation, answered a lot of questions about the system, and then suggested we meet up at Spondivits for a few beers. It was literally walking distance from the Delta facility. Carter knew exactly where it was, and asked if it was OK to bring along another good contact for me from Delta.

No brainer…yes, absolutely!

I arrived at Spondivits, and soon Carter showed up with BJ Smith, one of their Delta Program Managers, who would also be involved in the new B737 acquisition project. We ordered my recommended bucket of shrimp and a huge pitcher of beer…ok two…maybe it was three. What a hoot—these folks were really fun. I actually hurt myself laughing at some of their jokes, especially BJ's.

We chatted late into the evening about their US Navy and Air Force backgrounds, the roles they played at Delta, and why I thought the HGS was good for their new aircraft order. These guys were both familiar with the technology, since their US government fighter aircraft had all been HUD-equipped.

They were very curious about the Southwest and Alaska decisions to go with the HGS on planes that already had autoland, and about how we had convinced Boeing to agree to factory installation. They laughed at some of my stories and sympathized with the hardships.

A few days later, back in Oregon, I told Dick about my discussions with Delta, and their intention to purchase a new fleet of 737NGs. I said that I thought this airline might be possible as a customer, but I needed backup. Dick was more than willing. I called Carter later that week and suggested another "bucket party" at Spondivits, to introduce Dick. Carter readily agreed, and promised to bring along another good contact from their organization. The next week, Dick and I flew to Atlanta and drove to "my" restaurant. Carter came along with BJ, and a new fellow he introduced as Greg Saylor, a Delta Project Pilot.

These guys were all ex-military pilots so Dick, with his USAF background, made instant good friends with them. We spent an entire great evening over beer and shrimp buckets and a helluva collection of military pilot jokes and stories. Dick was able to explain some of the HGS benefits that I, as a non-pilot, could not. It was obvious we were to become a real team.

These were amazing get-togethers that would again change aviation on a huge scale.

A short while later, Greg told us that he, Carter, and BJ had briefed their Flight Ops management, and that he had been given the HGS project as one of his primary responsibilities at Delta. He would need to keep things alive with various other groups inside Delta, some of which were not as excited as our buddies were with the concept.

After our initial M-Cab session in Seattle for the Delta group, we were back in Atlanta. At another bucket party "meeting," Carter brought up a new subject.

"Phil and Dick, as you both know, we absolutely love your HGS," he began. "But your combiner! It looks like it was designed by the Russians in the 1950s."

"What? You gotta be kidding!" Dick and I answered in unison.

A chuckling BJ and Greg nodded in agreement with Carter's comments.

"What do you mean...what do you want it to look like?" I asked.

"Not sure, but it needs to look a little more modern for our new B737NGs," Carter replied, and again, both of his colleagues were nodding.

Dick and I knew we would have to bring this input back to the plant, and were unsure how it would go over with the A-Team.

After we returned to Oregon, we called Bob Wood into John's office to relay the "Russian design" comments. He was offended, but I had come to agree with the Delta assessment. The "Russian" design was OK for the B727 round dial aircraft, but for a brand new B737NG with a full glass cockpit, we needed something new. I told Bob it had to look "sexy." Bob, looking quite huffed, suggested that I come up with some "sexy ideas" myself.

I thought about this for a while. One day I met up with Bob in our cafeteria. "Why can't we make the combiner in the shape of a TV screen, curved on all sides?" I asked.

That got Bob thinking, and he agreed that maybe it was time to come up with a newer and more modern combiner design for our next generation HGS. He suggested he might also modernize the overhead projector unit as well. Great!

Over the following weeks and months, Bob began serious work on a new combiner design. Influenced by the competition from Europe and the new high-efficiency combiner coatings, he had already given up on the embedded hologram sandwich concept for the combiner glass and was using a single piece of thinner glass. At the same time, the Flight Dynamics design and manufacturing folks were transitioning to the computer age. Norman Jee had been working with Ziba Designs in Portland to get some industrial design inputs. All these changes came together about the same time as Carter's "Russian combiner" comment. Thinner glass, machines that were no longer limited to straight cuts, and CAD design tools led to a new, "sexy" combiner design.

All this time, Dick and I stuck like glue to the Delta team, with more meetings and presentations to Delta's various groups, and more after-hours bucket parties. We also invited more of their management to the M-Cab in Seattle for simulator sessions, and even to our facility in Oregon, where we arranged some good meals at great seafood restaurants in the Portland area.

At one memorable dinner at Salty's on the Columbia River, BJ had our whole group howling in laughter at his jokes. We were almost thrown out of the restaurant for disturbing other diners. Delta had brought along their FAA guy who—leaning way back in his chair, hands over his face, tears of laughter streaming down his cheeks, his chair wobbling—almost fell out the open restaurant window into the river below.

At our facility the next day, we decided to show the Delta team our new, sexy combiner design concepts, and they were thrilled.

The B727 system had been the HGS-1000, and the HGS-2000 was for the Dash 8. The original B737-300/500 was the HGS-2350, and the 3000 series was for some of our budding bizjet OEM customers, like Dassault. So the sexy new B737NG system for Delta inherited the next available number. It became the HGS-4000.

On a later trip to Atlanta for still another bucket party, Carter told us we needed to meet Bill Watts, his boss and their Director of Flight Ops Technical. OK!

The next day we met with Bill, who we also liked. He had been thoroughly briefed by his guys, and was receptive to Dick's and my executive version of HGS-101.

We later got Bill into a simulator session at the M-Cab and, of course, he was duly impressed. Upon his return to Atlanta, he became a strong advocate for the HGS. Delta was now in the throes of final negotiations with Boeing for their huge order of B737NGs, and Bill was "greasing the skids" at the higher levels inside the airline. With the HGS now offered by Boeing as a customer option, it was easy for Delta to at least ask about the system for this new order. Interestingly, the system payback issue did not really surface as a major requirement at Delta—they were gung ho on the HGS primarily for its many safety benefits.

As all this was going on, our A-Team had not been idle. They had been looking into reducing the HGS takeoff minimums from 600 RVR down to 300 RVR, and had applied to the FAA for approval. This required a lot of support from the FAA, but finally certification was achieved, and this bolstered our payback analysis a lot. The 300 RVR takeoff was very appealing to Delta's flight ops group, since neither the aircraft's autoland nor any other aircraft system on board could offer the same operational capability.

Delta understood the CATIIIa autoland system, and precisely what was required for it to work. Not everyone there understood why they would also want HGS. BJ had come up with a new and improved argument for the HGS on an autoland-equipped aircraft. He considered it an "experience compensator." That is, the system allowed a relatively inexperienced pilot to fly to the same flight performance levels as a well-seasoned pilot. This became a huge flight crew safety argument in favor of our HGS at Delta.

Autoland plus HGS was a novel concept, and it found some support at a number of levels within Boeing. At this time, there was industry controversy over the high level of automatics on the Airbus A320. Some in aviation felt that automation had gone too far and could actually become a safety concern. Our manually flown HGS addressed this issue.

But a new wrinkle suddenly appeared. Actually, two...

Carter called me one day to announce that a potential competitor had shown up—Sextant-Avionique, the French HUD manufacturer. We had heard a little about them when we were chasing Northwest Airlines. Apparently, they had been successful in achieving CATIIIb approval from the European JAA—the FAA equivalent "across the pond"—on an Aeropostale B737-300, using their HUD and the aircraft's autoland. They were now scheduled to visit Delta the following week to provide a presentation and demonstration of their HUD system. I suggested to Carter that Dick and I could meet him for a bucket party after the demo at our favorite spot. He agreed.

Dick and I flew to Atlanta and waited patiently at Spondivits, trying to envision what was going on and what advantages the competitor might have over us. We did not think Delta was set on CATIIIb, but we were in the dark.

Sextant had originally been called CSF Thomson, and had done some other commercial HUD work in Europe on the Dassault Mercure aircraft. There was speculation that they were now proposing the same hybrid system approach to Alitalia for their MD-80s. But we had seen no evidence of any more recent activities, especially for the Boeing B737NG.

Finally, our smiling Delta team showed up. After ordering the standard bucket and beer pitchers, we peppered them with questions about our competitor. The more beer we drank, the more we listened...and were relieved to hear that we were in good hands—our Delta team were still strong HGS fans.

One thing came up that was quite funny to all of us. On our HGS, as the aircraft went into the flare movement—the gentle nose-up maneuver to ease the landing impact—our system displayed an "Idle" command on the combiner, which instructed the pilot to bring the throttles to the idle position to ensure precise touchdown. For the same scenario, the French HUD system displayed the term "Retard," meaning to retard the throttles. The Delta folks had flown the Sextant-Avionique system on an Airbus A320 simulator while evaluating the B737NG, and at that time they had explained to Sextant that the term "retard" had another, not so flattering, connotation in the US. They had asked if the French could use the word "idle" on their HUD system also.

"Non, monsieur, le mot correct est retard," was the reply, imitated so well and hilariously by BJ. ("No, sir, the correct word is retard.") Delta was stunned—why not agree to such a simple change? The French resistance to a minor request worried Delta. If other required changes came up later, would they receive the same kind of rejection? What happened to "the customer is always right"?

On the day of the Delta demo, Sextant, which had no simulator asset like the M-Cab, had shown up at the Delta Flight Ops building with their "show simulator." It was very large and heavy. They had to rent a truck to get it from the airport to Delta's facility. And, once there, the only place it could be set up was near the elevator lobby on the fourth floor, since none of the rooms were big enough to accommodate the simulator plus the expected visitors. After many hours of setting it up, the Sextant folks were unable to get it running properly to provide any HUD demos to the Delta team or their management there.

This episode, though unfortunate for the Sextant folks, increased our chances of winning the sale.

A short while later, during a catch-up phone call, Carter told me that yet another competitor had shown up and wanted to present their wares. This was GEC Marconi, the UK company we had heard about previously during their partnership with Honeywell for the Gulfstream bizjets. GEC had a lot of HUD experience—they had designed and built over sixteen thousand HUDs, a large majority of those for the F16 fighter. However, while they had beaten us to the Gulfstream bizjet program, and had done well with their HUD on that aircraft, they had no commercial air transport HUD experience.

Greg was told by Bill Watts to keep the competition doors open. But with no air transport aircraft FAA certification experience with their HUD, GEC was perceived at Delta as a bit of a risk. Greg was convinced that the HUD they picked had to work properly right out of the box. Any early teething problems could have a catastrophic result on their entire program.

Flight Dynamics was continuing to hold annual HGS Operator Conferences at some of our newest customer locations. All current customers were also invited. The next conference was slated to be at the Southfork Ranch near Southwest Airlines headquarters in Texas. The facility had been made famous by Dallas, a popular TV show at that time. We had picked this spot because of the HGS commitment from Southwest. I suggested to John and our marketing team that we invite Delta, even though they had not yet signed up.

Once again, he thought I was nuts. John seemed to often wonder about my mental stability!

"What if Southwest or Alaska bring up some HGS issues?" he demanded.

"Do you really think Delta will not find out?" I asked him. "Better to hear directly from us how we plan to solve any issues that do come up."

Dick was on my side, and John finally agreed. We invited our Delta buddies to the Southfork event and they were thrilled. It turned out to be a great move, allowing our Delta team folks, now sporting ten-gallon cowboy hats, to meet some of the Southwest and Alaska HGS advocates. It provided them many opportunities to ask operational questions and discuss training, simulators, and other relevant airline topics. The event turned out to be a huge success.

Carter and Greg finally told us that GEC's lack of experience with the development and FAA certification of a commercial transport HUD would bode well for our HGS. (Gulfstream had been the certification authority for the HUD on their bizjets, and Honeywell had provided the complex HUD computer.)

Throughout this process, Greg had to fend off many naysayers in Delta's engineering group, called Tech Ops. Some of them suggested that the HGS was merely a new pilot's toy, and they had to be convinced of its "across the board" benefits. Greg did a superb job.

We had spent an inordinate amount of time educating the different disciplines at Delta, including a buyer, John Buchanan, who had been assigned to work with us on pricing and sorting out any contract terms and conditions. Dick and I had met with him a few times while in Atlanta, and explained the HGS in layman's terms. Although we often invited him, he was never able to join our bucket parties.

One day, back at the plant in Oregon, not long after the Southfork event, I was instructed to come to John Desmond's office immediately and to bring Dick along. John explained that Buchanan was about to call with some final negotiation items, and he wanted us there to help sort them out. John's phone rang. On the squawk box, the discussion went on for quite a while, with us giving here and there, until Buchanan said, "OK, I guess we're done."

John continued to discuss where else we could give a little. Dick and I immediately jumped up and began slashing our necks, like, it's done, we can stop.

"Does that mean we have a deal," I asked Buchanan.

"Yes, it does. We have a deal," he replied.

We had done it! This contract would be for 300 HGS plus some spares and some options for even more new aircraft from Boeing. Once again, it was time to celebrate a huge win as we raced off to our favorite drinking hole.

This win was to REALLY get Boeing's attention.

Bucket party at Spondivits (l to r, facing camera) Greg Saylor, Dick Hansen, Carter Chapman, and Chris

HEAD-UP DISPLAYS

The pilot of an L'Aeropostale 737-300QC is using a Sextant Avionique Head-up Flight Display System for guidance during an approach in low visibility.

Sextant HFDS also is certificated in the Airbus A319 and is planned for certification in the Bombardier Global Express business jet and de Havilland Dash 8-400 commuter turboprop.

Although the Honeywell/GEC HUD 2020 is new to the market and first customers have not gained substantial operational experience, it has won significant market approval. About half of all new Gulfstream 4 and G5 customers have specified it for a total of more than 50 orders.

FAA-approved to Cat. 2 landings in the G4 and G4SP, certification of the HUD 2020 in the G5 is scheduled for this month. The $400,000 system features a "symbolic runway" that pops up on the display and accurately indicates the runway edges during approach. This is a big advantage during low-visibility conditions, according to Bob Morris, Gulfstream HUD program manager.

The HUD 2020 has a 30 X 25-deg.

The Sextant-Avionique HUD on an Aeropostale B737

A new Delta Air Lines Boeing B737NG

Captain Carter Chapman flying the new "sexy" HGS-4000

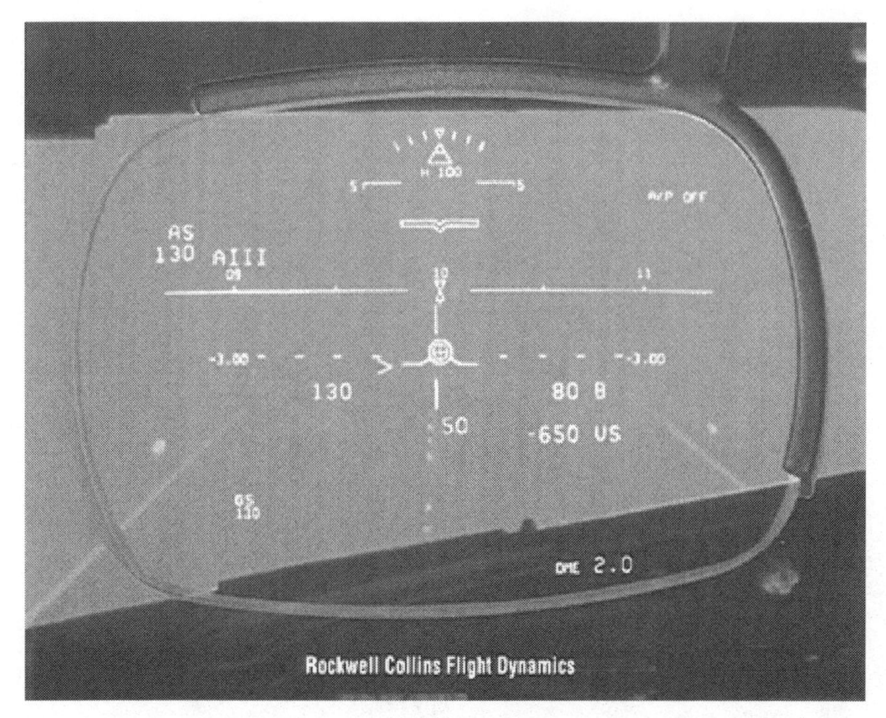

The HGS-4000

Our HGS Operator Conference at Southfork Ranch in Dallas, Texas. Getting bigger every year!

Chris and me at Southfork Ranch for the HGS Operator Conference

Chapter 14

Another Biggie

One day in 1996, John Desmond called me into his office to discuss something. He was intrigued by a little company in Sugar Grove, Illinois, not far from Chicago. They had developed a small HUD and had actually made some retrofit sales on older Gulfstream bizjets and King Air turboprops. The company was called Flight Visions. John wanted me to visit them to assess whether they might be a good acquisition for our company. He had confirmed that Collins-Kaiser would support this.

I did some research and found out that the president of Flight Visions was a fellow named Bob Atac. I called and the receptionist connected me to Bob's office. He was extremely cautious at first. I didn't blame him one bit. Eventually, after I'd explained what we were after, he relented and we set a date for me to visit.

There I also met Brendon Docherty, their Director of Sales & Marketing. Bob and Brendon were cordial and polite, and complementary about Flight Dynamics' successes, especially regarding progress at Boeing. They'd obviously had some interaction with that OEM, but we did not get into it.

It was the first time I had been inside another HUD company to see their products, and it was quite interesting. Their primary HUD was called the FV-2000. They took me on a tour but were careful to omit any key elements of their product technology and sales strategy. Their FV-2000 was certified on some aircraft, but not for CATIIIa or any poor visibility credit—it simply displayed flight instrument information without the pilot having to look down. But it had been approved as a primary flight display—no small feat—and had many of the same safety features we were touting, such as flight path vector, acceleration caret, and head-up operations.

Back in Oregon we had often talked about a "baby HUD" for smaller aircraft, and I think John thought it might be easier to acquire this company than to develop our own small HUD. I spent the entire day with Bob and Brendon, including taking them to lunch so we could continue our discussions. I mentioned that we were impressed with their company's retrofit HUD business and successes. More than anyone, we could appreciate the effort that had gone into it.

I returned to our office, where John, Jim, and Bob Wood were anxious to hear my opinions about what I had seen and heard at Flight Visions. I told them that I thought they had no "secret sauce" when it came to HUD technology, but they did have a lot of able-bodied HUD engineers and manufacturing staff that we could develop into a small Flight Dynamics HUD subsidiary to target the smaller bizjets and turboprop aircraft. I was disappointed that it never went anywhere. Later, Flight Visions was acquired by the Canadian Marconi Company (CMC), based in Montreal.

During Delta's many visits to the Boeing M-Cab, our buddy Carter Chapman had started thinking that they would like to have AoA, or Angle of Attack, displayed on our HGS. This was the angle between the centerline of the aircraft and the flight path vector. Carter's colleagues completely agreed.

This request created a huge issue at Boeing. The Flight Test group there said that AoA was not present on the aircraft's head-down displays and they were absolutely against including it on the HGS.

But Carter and BJ insisted that they, as well as any ex-military fighter pilots, were used to flying with AoA displayed on their military aircraft HUDs, so why not have it on a commercial aircraft HUD? Dick, having also flown military fighters in the USAF, totally agreed. But Boeing continued to resist.

Finally, Delta came up with a great argument. If AoA was unimportant, why was it included in all flight data recorders that were used to investigate any aircraft incidents and accidents? That was a good question, to which Boeing did not have a good answer. Our A-Team was once again called upon to innovate and keep Delta happy, and the AoA was added to our HGS-4000 symbology display.

While I was busily engaged in the Delta HGS campaign, the aerospace industry was going through some changes.

Many aviation business jet OEMs were getting more interested in Enhanced Vision Systems, or EVS. This technology uses an infrared camera to detect objects and terrain in low-visibility conditions—much like night vision goggles. The scene is projected onto the HGS or any video-capable HUD. This was not a new idea—most military HUD-equipped fighters had an infrared image added to the HUD. The objective in commercial planes was to achieve still lower landing minima. EVS was being seriously investigated by Gulfstream in the mid-90s, using their GEC-Honeywell HUD and a Kollsman EVS. FedEx was also looking into EVS, and referred to the HUD-EVS combination as the "Magic Window."

As a result, Boeing became a little more interested in the potential combination of EVS and HGS. They created a project called ESAS, or Enhanced Situation Awareness System, to explore the benefits, technologies, suppliers, and potential certification of EVS. That led to some sensor manufacturers contacting our little company to explore possible joint business opportunities and relationships. One of these was Westinghouse. Their Seattle rep was Tom Kilbane.

I met up with Tom one day in Seattle and enjoyed chatting with him about our backgrounds, our companies, the Boeing ESAS project, and what each organization brought to the party.

Meanwhile, some personnel changes were taking place at Flight Dynamics.

Back at the plant one day, I found out that Jim McKinney, our newest marketing guy, had just resigned. In a meeting with John to discuss what to do about it, I recommended he meet with Tom Kilbane to see if he might be a good candidate to replace McKinney.

I probably should have asked Tom first, but I called him right away, with all my persuasive arguments at the ready, to ask if he was even mildly interested. To my surprise, he answered with a resounding "yes." On the day of the interview, I greeted Tom in the lobby of our facility and took him up to John's office. I stayed for the intro and some small talk and then left the two of them to sort out the details. About two hours later, Tom came by my office to announce that John had hired him. We decided to celebrate with a multi-beer get-together after work at one of our favorite restaurants in the area, a McMenamins Tavern called John Barleycorns. There, I was able to provide Tom with some additional info about the company and the HGS, and introduce him to George Kanellis and Tom Geiger.

Kilbane was single, with no kids, and willing to be on the road a lot. He would be looking after Europe for us, and lots of travel would be involved.

Right on the heels of Jim McKinney's departure, Jim Gooden decided it was time to retire, and he submitted his resignation letter to John.

I was stunned...

Jim had guided me through the thick and thin of selling frozen turkey heads, and his leaving was a personal loss for me. I tried hard to dissuade him, but he felt that the company was now on a good track, and he was determined. At his age, he wanted to take life a little easier. Who could blame him? But he asked me to stay in touch, and to update him on all the latest Flight Dynamics HGS activities. We agreed to a call near the end of each month, where I would provide him the latest on all the happenings at our little company. He was always thrilled to hear about us moving forward with his frozen turkey head campaigns.

John accepted Jim's resignation and began the hunt for a new Director of Sales and Marketing. I was not remotely interested in the position, since it required a lot of in-office time and effort with our management as well as our owners, and my interest was in the customer chase. Our new owners suggested Marc McGowan, who was a VP of Sales at Honeywell in Phoenix at the time. Scarcely a couple of weeks later, Marc showed up in Portland for an interview with John. They had a good meeting, and Marc accepted the position.

We had another celebratory meeting at Barleycorns to introduce Marc to everyone. We now had two high-caliber additions to our sales and marketing team!

A little later, I was sitting at my office desk when my phone rang. It was one of our marketing buddies from Bombardier, Rod Sheridan. He told me that Bombardier had made a deal with Horizon Air to bring one of their HGS-equipped Dash 8 aircraft to the Farnborough Airshow in the UK. This was a huge global show, and we would get lots of industry exposure. We were thrilled.

Rod asked if one or two Flight Dynamics folks could support the Horizon personnel at the show to answer customer questions about our system. Since Tom Kilbane now handled Europe for us, he was nominated. John, who also planned to attend, suggested that I come along to assist Tom. Sure, why not?

Tom and I spent quite a bit of time at the Horizon aircraft display answering many HGS questions from interested parties. This was indeed a great opportunity for us.

But we did have some time off. Since I covered Boeing for our company, I had contacted a number of folks I thought might be at the airshow and received multiple invitations to come and visit them at the Boeing chalet. While he'd been with Westinghouse in Seattle, Tom had also made some good inroads at Boeing, so between the two of us, we had many people to see and update on the HGS.

So one day Tom and I headed over to the Boeing chalet. John was busy with some Collins folks, but the Horizon fellow felt comfortable by now handling any HGS questions.

We managed to get past the "chalet Nazis" and quickly found a couple of our Boeing contacts. This was a great opportunity for us to meet some of the senior Boeing executives, and to introduce ourselves, our company, and the HGS. Tom and I knew some of the same folks, but we introduced each other to many others, making the visit that much more valuable. We enjoyed a nice lunch and a few hours with our buddies.

One of our parent companies had a couple of very robust gentlemen in their sales organization. As Tom and I were leaving the Boeing chalet, these two big guys happened to be walking past. They cornered us and asked what the hell we were doing there, and where was our president, John Desmond. We explained the purpose of our meetings, and told them we had no idea where John was. They were shocked that we had been into the Boeing chalet without "adult supervision."

We laughed ourselves silly later, but it was to have a lasting and quite negative personal impact on me.

Another funny thing happened about this time. It occurred during another fog bout at Seattle airport. A brand new United B777, which had been delivered only a couple of weeks earlier, was getting ready for its first revenue flight to London, England. It had pulled away from the gate with a full load of passengers and was taxiing toward the takeoff runway position when the pilots were asked to hold up. Air Traffic Control (ATC) advised them that the visibility was now only 400 RVR, and that was below their takeoff minima of 600 RVR.

After waiting for about forty-five minutes, the B777 flight crew was asked by ATC controllers to move onto a nearby apron to allow another aircraft to pass. The pilot asked if the RVR had improved. ATC said no. Incredulous, the pilot asked what was happening. ATC said there was an aircraft behind them that was cleared for 300 RVR takeoff. United asked how that was even possible and ATC said the aircraft was HGS-equipped. As the United crew watched in dismay, a Horizon Dash 8 bound for Portland taxied past their new, and now idle, B777 and lined up for takeoff. They could just barely make out the small plane in the dense fog.

Later, the CEO of United wrote a letter to Boeing senior management lamenting that "it was a sad day when a small regional aircraft like the Dash 8 could outperform Boeing's Queen of the Sky."

What a hoot!

In the meantime, American Airlines came alive, and began asking questions about why Delta had selected the HGS for their new fleet of CATIIIa-capable B737NGs. According to the American Airlines customer engineer at Boeing, the airline was looking to replace a very large fleet of two hundred and eighty older MD80s and a hundred Fokker F100s with the newer B737NG model. One of our Delta comrades provided me a name at American Flight Ops in Dallas: Captain Brian Will, a Technical Pilot who was also on the B737 acquisition team. After a good chat on the phone, he asked if it was possible for me to come to their headquarters in Texas for a full-blown presentation and some initial HGS discussions.

A week later, I arrived at American's facility near DFW airport and asked for Brian. He arrived in the lobby and escorted me up to his office. After an introductory chat, he summoned another comrade, Captain Rick Owen, who was also on the B737 acquisition team. I found myself liking both of these two pilot types. I took them through my HGS presentation and answered many of their technical questions, especially about the redundant CATIIIa capability. I focused on the "experience compensator" argument that BJ had used so successfully to help push the project at Delta over the top.

American Airlines had experienced a fatal B757 accident in Cali, Columbia, in 1995. The cause was an incorrect entry in the aircraft's flight management system, or FMS. As a result, the airline's flight safety and operations groups had become concerned that maybe there was too much flight crew reliance on the aircraft's automation systems. While looking into the B737, they had also considered the Airbus A320, a highly automated aircraft. Flying the Cali accident in the A320 simulator reinforced their concern that there was too much automation in the cockpit. Pilots could be heard saying, "What's it doing now?"

Because the new B737 fleet would be flown by entry level pilots at American, the automation concern became a much bigger issue. The airline believed that these entry-level pilots should be mastering their "stick and rudder" skills instead of just pushing buttons. Manually flying CATIIIa with HGS provided the safety advantage they were looking for, and helped bolster the rationale for selecting our system on a CATIIIa autoland-equipped aircraft.

Brian and Rick were ex-USAF pilots, so Dick's background once again was an advantage. Dick worked his magic at a few M-Cab sessions with both pilots—including flying the Cali accident simulation.

Dick and I then visited their headquarters at DFW to meet some other American employees, including Paul Railsback, a Flight Ops Manager, and Captain Warren van der Burg. As a result of Cali, Warren had created an unusual attitude recovery training program for the pilots at American. He became quite fascinated with the attitude recovery symbology that Dick had helped to develop for our HGS.

We also met the VP of Flight Ops, a Captain Cecil Ewell, who we knew would become more instrumental as the project moved up the corporate ladder. Cecil was great! He allowed me to pop into his office unannounced when I was in town catching up with Brian and Rick. Even his secretary knew not to try to stop me. He was always eager to hear the latest news about our system, and any potential customers that were banging on our door. He was fascinated by what was going on at Boeing, and extremely complimentary about us getting this OEM to agree to offer our system on their new B737 as a factory option. I emphasized the gigantic Southwest assist, and the importance of big airline deals in getting concessions from the Big B.

One new thing emerged, and it was to become a major problem.

American and Delta were competitors, and even though at the level we were working it did not seem to be a problem, there was still some awkwardness.

On a subsequent trip to American's headquarters, Dick and I noticed three names on the log book sign-in page in their lobby. They were from GEC Marconi, the UK company we had encountered during the Delta and Gulfstream campaigns.

Uh-oh, were we squaring up for another HUD competition?

We became quite worried that, having lost the Delta HUD program, GEC would now be far more aggressive, and trying even harder this time around. They had since managed to get their HUD FAA certified for CATIIIa on a leased B737.

Although they were big in HUDs for military fighters like the F16, and had delivered thousands of HUDs for that aircraft alone, they'd still had no experience getting into the commercial air transport HUD marketplace. They had however, managed to win the FBW (Fly By Wire) system on the new B777 wide-body aircraft now in production, and that gave them a leg up with engineering and senior management at Boeing.

Soon after arriving for a subsequent meeting with the American team, Dick and I peppered Brian and Rick about the GEC visitors. They admitted that they were going to get HUD proposals from both companies for their new Boeing airplane order. GEC were calling their HUD the Visual Guidance System, or VGS.

I did not like the sound of this at all!

The Flight Visions FV-2000 Head-Up Display wasn't really serious competition to the HGS.

Dick Hansen, Chris, Jean Hansen, Therese and Marc McGowan, and me

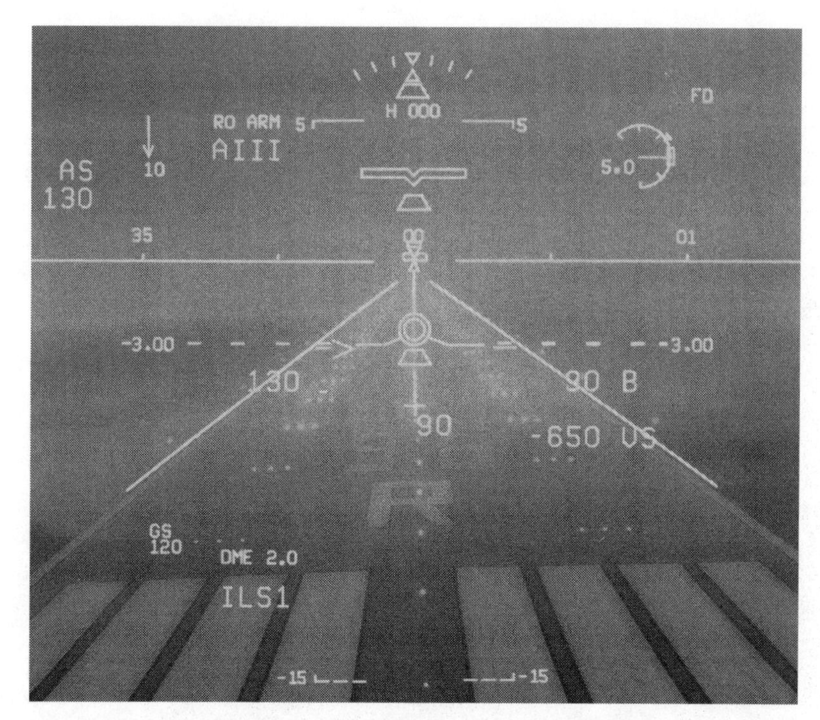

Finally, we got AoA presented on the HGS

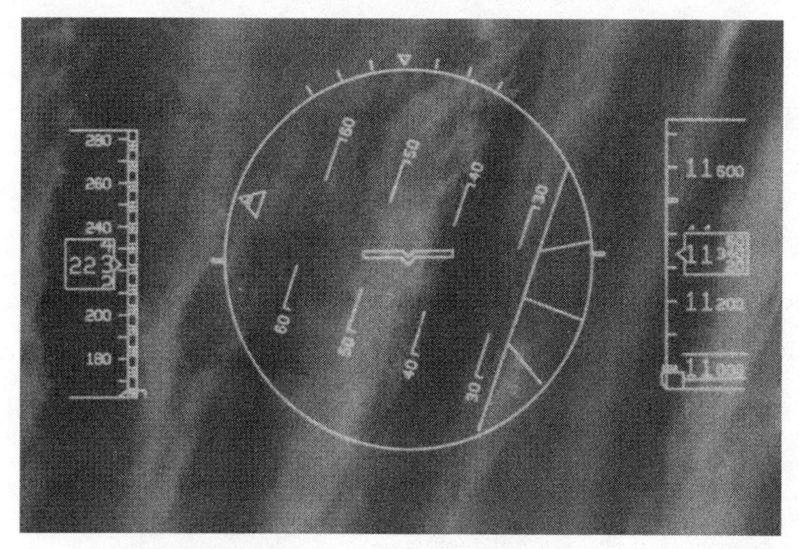

...along with unusual attitude recovery

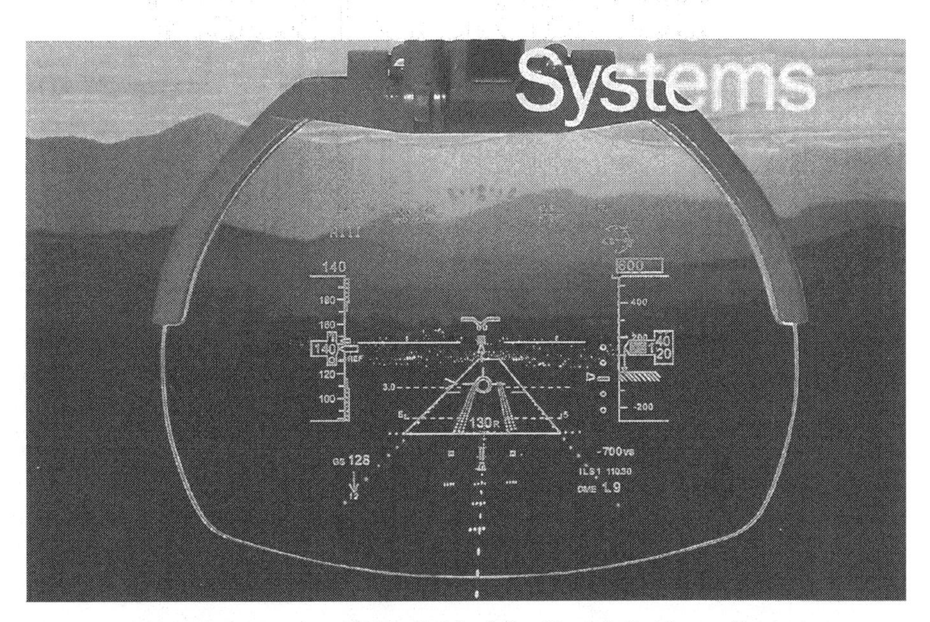

The GEC Marconi HUD, dubbed the Visual Guidance System

Tom Kilbane, Joe Marott of Southwest, and me celebrating at the Collins chalet in Farnborough, UK

HGS-equipped Dash 8 at the Farnborough Airshow

Tom Kilbane with one of the Horizon folks at the Farnborough Airshow

Chapter 15

An Ugly Christmas

At our facility in Tualatin, Oregon, we were now having ongoing, serious, and often heated war room discussions about American Airlines and our emerging competitor, GEC Marconi. Without a doubt, GEC was going to be a viable and strong competitor, and did not suffer from the same self-defeating philosophy as Sextant-Avionique. The newest addition to our marketing team, Marc McGowan, was a big help, since he had faced many competitive scenarios just like this one in his career at Honeywell. He was able to share some interesting strategy and tactics with our management and our marketing team.

One big advantage that GEC was touting was their "Mounting Tray" approach. They had come up with this solution to minimize design changes as the VGS was adapted to various types of aircraft. They employed a tray of sorts, and their VGS attached to the bottom of the tray. Only the upper portion of the tray needed to be altered to suit other aircraft—a far less expensive solution than the complete redesign that our HGS required. This was a big factor in American's assessment of the two HUD choices, since they might also want the system on their large B757 and B767 fleets. This significant competitive advantage worried me…a lot!

On a summer 1997 trip to Dallas with my wife, Chris, for yet another LUV Classic barbecue and "Steakout" with our Southwest buddies, American's Brian Will invited Chris and me out on his new powerboat, an impressive 47-foot Maxim, built in Oregon. What a yacht! He had gotten a great deal on it by agreeing to allow it to be demonstrated by the local Maxim dealer to other boat buyers in the area. Onboard, we met his wife Martha—dubbed "The Major" by Brian—and his two young sons. We had a great day on a large lake not far from their home (although I went pale watching the marina gasoline dials as he filled the boat before departing). Chris and I invited them to dinner afterward. While Brian and I touched slightly on the current HGS campaign, we did not get into too many details with the families present.

Back in Portland, Flight Dynamics received the formal HUD Request For Proposal, or RFP, from the folks at American. After the A-Team reviewed our proposal and gave their input, I spent many of the following days, evenings, and even some weekends preparing, updating, and modifying our document. After getting John, Marc, and the Collins folks to review it and suggest changes, and after many iterations, we finally agreed that what we had was the best we could do. I submitted our effort...

On our next regular trip to Dallas, Dick and I were once again working late. As I went down the hall to find the men's room, I happened to glance through the window of one of the offices and saw the GEC technical proposal sitting on a desk. The door was open so I popped in to take a look... I managed to get a glimpse at some of the technical stuff, but it was not immediately apparent how any of what I was looking at would be of value to us.

For the next couple of weeks we waited...and waited. It seemed like an eternity. I called Brian and Rick on a regular basis, but the answer was always the same: "We're still working on it."

Finally, the anticipated call came. It was from Brian.

"Phil, we've completed our in-house evaluation of your HGS proposal. While it was very thorough and professional, we still have many questions. And we have a lot of questions for your competitor also."

"OK. So now what?" I asked.

"We want both companies to come to Dallas for some final discussions next week. Can you arrange for your team to be here then?" he inquired.

"I...I guess so," I stammered.

This was a bit weird, to say the least. Two fierce competitors at the customer's facility at the same time to negotiate a final deal.

I immediately met with our A-Team, which now included Marc, and spilled the news. Silence. It had been relatively easy to overtake Sextant-Avionique and GEC Marconi during the Delta campaign, thanks to lots of good discussions and information exchanges at our bucket parties. But the more competitive nature of American would not provide us the same opportunity.

Our team was all set—Marc, Dick, and I would head to Dallas the following Monday, December 15, and meet with American at their Flight Ops building on Tuesday. We stayed at the Holiday Inn North, close to the airport and the American facility. When we arrived early Tuesday morning, Brian greeted us and led us upstairs to a small conference room with about eight chairs. Coffee was brought in—a big jug.

"Looks like we're gonna be here for a while," I said, pointing to the mammoth container.

Marc, while looking around the hallway outside our room, noticed that the GEC folks were in another room just down the hall. Very strange indeed! We recognized Ron Barry, the Seattle GEC rep, and Glen Hislop, a technical type, pacing around with a couple of others, we assumed from the UK.

Over the next couple of long and exhausting days, the American team would visit each supplier's room in turn to ask various technical and business questions about our proposals. We were getting weary. We told the airline folks that we had return flights for Friday evening. We wanted to be back in the Pacific Northwest for Christmas the following week. Each evening, we returned to our hotel and, over a late dinner and beers, we discussed the day's events and plotted changes to our strategy for the following day. But each day, new questions threw us for a loop. We believed that the questions we were being asked were based on answers they had received from GEC, and we were sure GEC were thinking the same thing about us.

One major issue that came up was the price that Boeing charges for factory installation. We all knew the number for the Southwest B737 Classics, after the great job done by Jim Von Der Linn and Bob George at Boeing. But we had no idea if Boeing would increase the installation charge for the newer B737NG. It would be an issue for Southwest, Delta, and Alaska, however, no one had yet asked Boeing if there would be a cost increase.

On Thursday, American hit us with some new pricing demands and additional concession requests. One included a clause that stipulated that we would have to cover the increase, if any, in the Boeing charge for HGS factory installation on the new B737NGs. We knew immediately that this would not float with John or our CFO back home, nor with our owners. We told the American team that this would be a huge problem.

I phoned Carter at Delta and asked him to request the Boeing HGS install price on the new B737NGs. He said they had already done that and were waiting for the answer. It was not coming soon, but he and the Delta team were convinced that Boeing would not up the charge.

As we returned to our car in the parking lot that evening, we glanced up and saw GEC and the American team through the conference room windows. It looked like things were getting heated. They were probably being asked for the same concession.

During our strategizing session back at the hotel, I told Marc and Dick we should tell American that we had postponed our return flights until the following Wednesday, which was Christmas Eve.

"Why would we do that?" Marc asked. "We're not really going to postpone, are we?"

"Of course not," I replied, "but American does not want to be here any more than we do next week, so close to Christmas, and they certainly don't want to be working this issue over the weekend." If their strategy was to pressure us into giving them an answer on Friday, in order to catch our flight, it would backfire. Both Dick and Marc immediately understood and fully agreed.

The next morning, back in our little room with the American team, I beamed at them. "Guess what, guys? We thought this might drag on a little, and we didn't want to rush you, so we moved our return flights to next Wednesday so we can work this all weekend and even into next week if needed."

The group on the other side of the table looked totally dismayed, and we felt that we had accomplished our objective. The American team now accelerated their efforts to complete their investigation of the two suppliers. By early afternoon on Friday, we were close to finishing up. We were still rejecting their demand that we cover any differences in the Boeing pricing for factory installation, and we felt sure that GEC was giving them the same answer.

I managed to corner Brian in a hallway after our meeting concluded, and asked him how we did. He was non-committal, and told me we would have to wait until after Christmas to find out. But his look worried me. Catching up with the rest of our team at the elevator, I told them I was not confident…at all!

As we headed to the airport, Marc suggested we take a good holiday break and come back refreshed in the new year. Dick and I agreed. But I knew I would not be enjoying my Christmas as much as usual.

On a return trip to American in January, I caught up with Brian and asked him again how things were going. Again he was non-committal, only telling me that GEC were making good headway. I was quite concerned that we were losing ground. We knew GEC were lowering their prices, but we could not do the same because we had a Most-Favored-Nation Clause with other B737 HGS customers. Our hands were tied.

While this was going on, I had attended yet another NBAA and renewed my friendship with Bill Schultz, Manager of Aircraft Operations for Corning, in New York. He had been one of the first visitors at my first NBAA, in New Orleans, where we'd exhibited our HUD for the first time (a lot of firsts!). He asked me if I could come to Horseheads, New York, to discuss a potential HGS deal.

Corning was in the market for four new bizjets, and were competing the newest Bombardier CL604 against the Dassault F2000 for the approximately eighty million dollar deal. Bill had told their Bombardier account manager, Ed Osgoode, that they were not interested in the Sextant-Avionique HUD option on that aircraft, and if they couldn't get the HGS on the 604s, they would go to Dassault, where the HGS was offered as a customer option. I soon received a call from Bill to tell me that Corning had selected the F2000 because they could get our HGS on that bizjet. Yay, another win! Bombardier was incredulous that Corning would make such a huge aircraft decision based on our system availability.

Meanwhile, we still didn't have an answer from American.

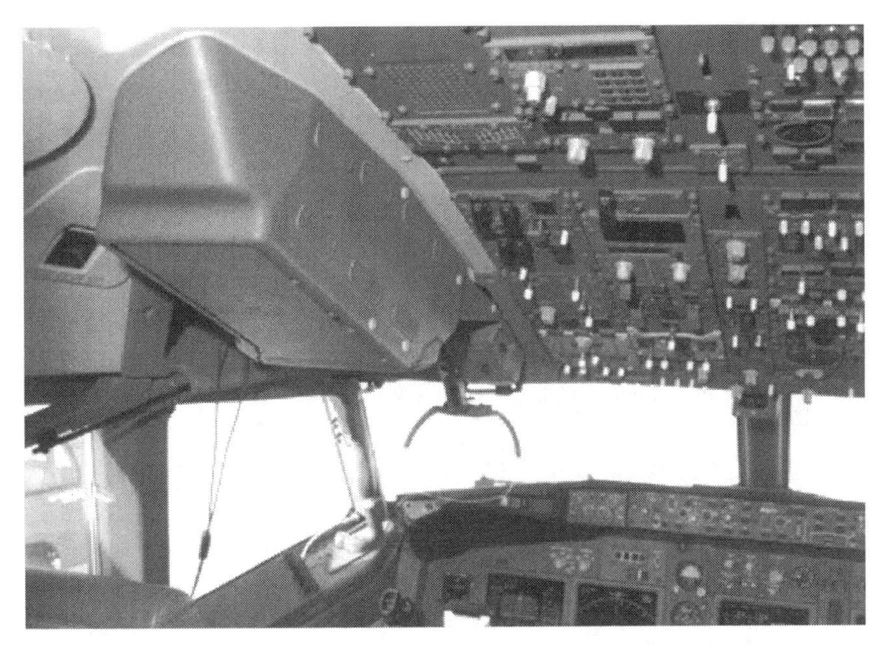

The GEC Marconi VGS installed in Boeing's M-Cab simulator

Chris, me, Brian Will, and "The Major" (Martha Will) out on Brian and Martha's new 47-foot Maxim

Brian Will, Chris, Brian's sons, me, and "The Major" after our boat ride

American Airlines' new Boeing B737NG

FARNBOROUGH '98

New GEC HUD Aims For Transport Market Growth

PAUL PROCTOR/SEATTLE

GEC is targeting the rapidly expanding airline market for head-up displays with an advanced, lightweight unit based on its military head-up display technology. A prototype is scheduled to fly this month in a Boeing 737-200 transport testbed from Mojave, Calif.

American Airlines signed as launch customer for the GEC civil head-up display in January, placing an order for 75 flight units plus simulator models for its fleet of new 737-800 transports. American holds options for up to 425 more. That total could grow even further if the Dallas-based carrier expands HUD use into heavier aircraft. GEC also is actively marketing the product to other airlines worldwide.

This *Aviation Week & Space Technology* editor recently performed a series of approaches using the GEC HUD installed in Boeing's M-CAB multipurpose engineering development simulator here. I previously have flown Flight Dynamics and Honeywell HUD2020 head-up displays in simulators and a business jet. France's Sextant Avionique and Flight Visions of Sugar Grove, Ill., also make HUD product lines.

HUDs display critical guidance cues, flight and aircraft system information on a transparent screen located between the cockpit windscreen and the pilot's eyes. This information, which generally duplicates cockpit panel displays, keeps the pilot's focus on the view outside the cockpit, without the need to break to scan the instruments and refocus. HUD's real-time information is especially useful during approach, landing and takeoff phases of flight, where a high percentage of all accidents occur.

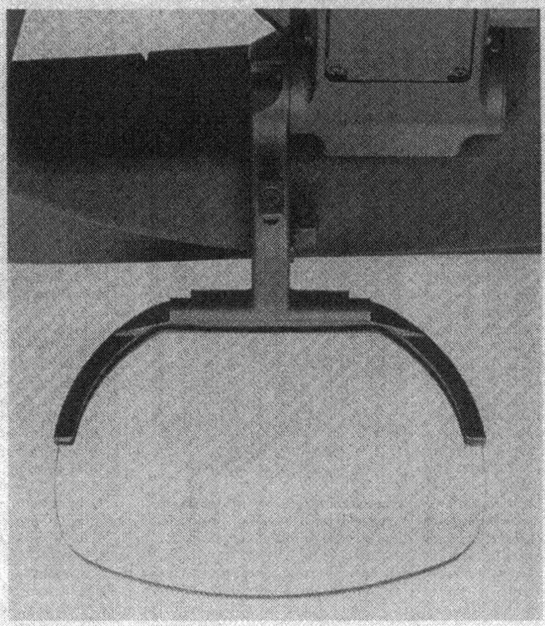

The GEC HUD's lightweight, 11-oz. screen eases installation and reduces the potential for pilot head injury in a hard landing.

ROCHESTER, ENGLAND-BASED GEC has integrated the best features of its many military head-up displays into an effective, versatile and lightweight package, according to Glen M. Hislop, U.S. program director for GEC's civil head-up display. GEC started work on a civil transport HUD almost two years ago, having delivered thousands of military HUDs worldwide. The civil HUD's technology includes an all-digital system and raster-capable display, assuring future growth capability.

With the world jet transport fleet scheduled to double by 2015, safety authorities and airline managements are interested in head-up displays as a means of obtaining the targeted 50% reduction in large jet transport accidents. In particular, HUDs offer flightcrews a means to increase safety margins during high-risk "black hole" and night and bad weather approaches to outlying airports with non-precision or Category 1 approaches, according to Brian Will, 737/MD-80 technical pilot for American.

Operational benefits include potential Category 3 qualification at outlying airports, lowering weather minimums and increasing dispatch and arrival reliability by allowing the pilot to monitor an automatic approach or hand-fly without frequent reference to the panel. The system also gives pilots a critical flight path cue in windshear and shows TCAS advisories as well as commands the proper flightpath angle for the escape maneuver.

Hud's precise, real-time touchdown guidance and related information also is expected to help reduce runway excursions, long touchdowns, tail strikes and hard landings. The latter two can be expensive for an airline even if no one is hurt, taking the aircraft out of revenue service for days or weeks for inspection or repair. "The expanded field-of-view of the HUD permits the pilot to hand-fly the airplane almost as precisely as the

148 AVIATION WEEK & SPACE TECHNOLOGY/SEPTEMBER 7, 1998

This GEC article came out in time for the 1998 Farnborough Airshow

Chapter 16

A Trip to Brazil

The American Airlines verdict finally came—GEC Marconi was selected as the winner in the B737NG HUD competition. While I had suspected this would be the outcome, it was still a terrible blow.

Almost at the same time, Marc McGowan decided to return to Phoenix, and handed in his resignation. His wife, Therese, was not at all enamored with the Portland weather. At one lovely summer barbecue in my back yard, with the temperature in the mid 80s, Therese had approached me.

"Hey Phil, when does it warm up here in Portland?" she asked.

"Are you kidding?" I replied. "This is as good as it gets!" Uh-oh…she did not look happy.

After a lengthy goodbye chat with Marc, he hinted to me as a parting suggestion that we had never spoken to America West, who were based in Phoenix. He offered to help however he could.

I called the airline to find out who I could talk to. I was given the name of Van Reavie, their Director of Safety. After an encouraging chat with Van, who had heard a lot of good things about our HGS, I decided to visit their facility near the Phoenix airport. There, I was introduced to Ed Methot, VP of Flight Ops, Randy Olin, their Chief Pilot, and Bruce Burnett, their financial justification guy. Bruce and I became good beer buddies.

At America West headquarters, I learned that the airline was operating thirty-eight B737-300s, the same aircraft we had installed the HGS on for Morris Air and a big portion of the Southwest fleet. America West was planning significant expansion into the Pacific Northwest and they had concerns about the frequency of fog.

The group had many questions after my HGS-101 presentation, mostly about Delta and Alaska, but also about why American Airlines had selected our competitor. I tried to address this last one the best I could, but focused on the fact that GEC had no certification on the older B737-300s. Given America West's competition with Southwest, and the probability of Seattle and Portland fog, they decided they should take a hard look at an HGS system.

Over the following month or so, Dick and I did our thing in the M-Cab with Van, Ed, and Randy, and they were all amazed. I was starting to feel good about the progress we were making.

At an Air Transport Association conference later that year, I ran into Ed. In a conversation over dinner one evening, he stressed that while America West was excited about our system, the airline had been through some recent tough financial times. Consequently, the airline's executives were now extremely careful about any major expenditures, and that would include our HGS. We would need to make a very solid business case. He told me he had assigned Bruce Burnett to support the preparation of our system's payback analysis.

Back at our plant, I met with Kanellis to update him on our progress. "America West in Phoenix is quite interested in our HGS but they will need a thorough payback analysis for a project go," I explained. I also mentioned the comment Ed had made, about the airline's financial situation.

"OK," he responded. "I have the NOAA data, but we will need their route system and flight schedule."

I got the information from Bruce and passed it on to George. Not long afterward, George slumped into my office with a long face, sat down, and sighed.

"Phil," he began, "we have a big problem with the America West economic payback. Their two main bases are Phoenix and Las Vegas. There's literally no fog occurrence at either of those cities, so the payback will look very marginal."

I asked him to process it anyway, but the final result was even worse than either of us suspected. I asked George if there was anything we could do to improve it and he said no, it would not pass the "magic two-year payback" hurdle that all airlines required. In fact, it was not even close, which probably explained why America West was only a CAT I airline.

I passed our payback analysis on to Bruce and Ed and was told they did not feel it was even worthwhile going up in front of the airline's executives to plead the case. We were toast. They had been quite excited about the HGS possibility, and were both disappointed in the results of our analysis.

After the ugly Christmas season we had spent working the failed American Airlines project, and the bad news outcome at America West, Captain Sunshine was a bit down...OK, maybe a lot. Two in a row! Marc's leaving was also a contributor. But I remembered that in the past, each time we'd faced a loss, a light had appeared somewhere on the horizon. This cheered me up.

As I was contemplating this, our receptionist called my office to say there was someone on the line from South America with some HGS questions, but she was having a hard time understanding him. Could I help? I took the call.

It was to be a good call!

It was a procurement fellow from Embraer, an aircraft manufacturer in Brazil, who mentioned that he had some engineering folks who wanted to chat with me. His name was Luis Antonio Junquiera, and he asked if it was okay to put me on speaker. Of course! I was suddenly talking to three additional people on a squawk box in Sao Jose dos Campos, Brazil, a small community not too far from Sao Paulo. Apparently, after watching the success of the fifty-passenger Canadair Regional Jet, Embraer had decided that they too were going to build a regional jet. They had come a long way, even getting their new EMB-145 aircraft certified. I had been monitoring their progress in some aviation magazines, so this was not a complete surprise to me.

However, they had a big problem. Their launch European customer, Regional Air, wanted CATIIIa, and Embraer had absolutely no autoland expertise or experience whatsoever. They knew our HGS was being offered on the CRJ and the Dash 8, so they were sniffing around our system as a potential solution. By the end of the call, I had promised to make a trip to South America to provide an HGS briefing to a much wider audience at their facility in Brazil.

I went to visit Desmond in his office and told him about my conversation.

"Are you crazy?" John asked me yet again. I shrugged to indicate the possibility.

"But why would we not be interested?" I asked. "It's an OEM, and we agreed long ago that the secret to our eventual success was with OEMs."

"Look, Phil, Canadair has saturated that marketplace with their regional jet, so how much room do you think there is for a second one?"

"Competition is always good for any marketplace," I responded, "and their aircraft will be less expensive than Canadair's."

John was dubious. He knew how few aircraft Embraer had sold up to that point.

"I promised them I would come and visit them in Brazil to present our solution," I continued.

"OK…but only one trip," he finally agreed.

"I can't do a deal with Embraer on a single visit!"

"Nope, only one." John wasn't budging.

Four weeks later, after getting my visa, I flew to Sao Paulo and arranged a driver service to take me to Sao Jose dos Campos, about forty-five miles to the northeast. It had been strongly recommended that I not rent a car and drive myself. Having never been to Brazil, I was horrified by the miles and miles of shanties and chaos I could see from the air on the approach to Sao Paulo. But Sao Jose dos Campos was smaller and cleaner, and seemed more prosperous.

After I arrived at the large and impressive facility and signed in, Luis Antonio, the fellow who had called me, picked me up in the lobby and walked me over to one of their engineering buildings. On the way, he told me to call him Antonio, since there were so many Luises in Brazil. We had a good chat on the walk and I felt like I would get along well with him.

I had a great audience of ten or twelve engineering and program management folks for the briefing, including the three folks I had spoken to on that first call, and I was received well by the whole group. They were convinced I had a potential solution for their European customer's aircraft. I could tell I was going to enjoy this.

I did my HGS thing, which took about an hour, but the barrage of encouraging questions and the detailed discussions afterward lasted another two hours. They were very interested, and were asking all the right questions. Antonio was already pushing on the rough price I had provided, hoping it would come down.

Two of the technical folks in the meeting, Rachel (pronounced Haakel) Penido and Adilson Antunes, both Systems Engineering Managers, were especially friendly, and offered to take me to their cafeteria for lunch after our meeting. There were many new-to-me foods and beverages. After lunch and an enjoyable conversation with my two new friends, Rachel told me I needed to meet one of their most senior people, Satoshi Yokota, who was a Senior VP and also the EMB-145 Program Manager for Embraer. He was the "grand fromage" who would make any final decisions.

We walked over to Satoshi's office. He was obviously very senior, as I spied a large and beautiful office that even included a big conference table. Beth Barco, his executive assistant, led us into Satoshi's office. Rachel introduced me, we sat down, and Satoshi got right to business. He really wanted the CATIIIa capability for their first European customer, Regional Air, but they were now also in initial discussions with Manx and Luxair, two other European carriers that were requesting the lower landing capability on their aircraft too.

Satoshi was serious but friendly, and he was eager to explore our solution. I took him through my presentation and rapidly realized that he immediately understood all the technical details and nuances, and I was able to move quickly through the overview. After inquiring about our rough order magnitude system prices, which he seemed to be OK with, he asked about our NRE, or Non-Recurring Engineering, costs for development and certification.

Our HGS would require certification from JAA—a European organization equivalent to our FAA in the US. I had discussed this with John before I left. We were looking at over five million dollars to complete a totally new HGS design and JAA certification for the Embraer aircraft.

When I mentioned the number to Satoshi, he went pale.

"Phil, we are talking about fifteen to twenty aircraft total, and I cannot amortize that level of NRE over such a small number," he replied.

We discussed whether Embraer could come up with a large portion of the NRE, but Satoshi explained that all their financial resources had gone into the EMB-145 development effort.

"OK, let me go back and chat with our management to see if we can find a solution," I responded, although I was at a loss to imagine how we could.

"I will need CAD files of the aircraft cockpit area," I added. This was a computer model that would allow our engineers to explore a good design fit for the HGS. Satoshi asked Rachel to provide me with the required files.

I waved goodbye to Satoshi and Rachel, and headed back to Sao Paulo for my long return flight. I thought about potential solutions all the way home, but came up empty-handed.

Back at the plant in Portland for our usual Monday afternoon marketing meeting, I brought up the discussion with Embraer and the issue of the NRE. John was adamant that our new owners would not accept investing that much NRE in this Brazilian aircraft for the small fleets involved. I struggled to convince him but to no avail. He agreed that we could look at modifying another HGS that we had developed for the Saab-2000. Based on the CAD file Rachel had sent to us, it appeared that only small changes might be required to modify that system to also fit the Embraer aircraft—this would reduce the NRE by about half. But the other system design changes and the European CATIIIa certification still amounted to a lot of money.

In a subsequent call, I told Rachel that we had found a design that looked like it might fit the EMB-145. She was ecstatic, and asked if we could send them a foam-core mockup—a plastic model, something many OEMs wanted—to check the fit and see what our proposed HGS might look like in the actual aircraft.

I asked John if that was OK, but he absolutely refused. In fact, he told me that without a firm commitment from Embraer to pay our total NRE, I was to "No Bid" the project. Period! I did not want to do that—it would be closing a door that I really wanted to keep open. But John was emphatic.

Here I decided to tread on shaky ground…again!

I went to Bob Wood. He called one of his designers over and asked if he could look at modifying the Saab 2000 HGS to fit the EMB-145. I was determined to find a solution that might minimize our NRE. Shortly afterward, Bob called me downstairs to take a look at the computer model. I was amazed at the result—it looked terrific. I asked how difficult it would be to build a foam-core mockup, and explained that I did not want any company hours used. Bob asked the engineer if he could do this in his spare time, and I offered to pay for a nice dinner for him and his wife. Deal!

Three days later, I walked into my office early and found a green garbage bag on my desk. Inside was the foam-core mockup, and it looked great. I hid the bag under my desk. At lunchtime, I exited out the back door of our facility with my prize, drove to a local DHL office, and shipped the mockup to Rachel in Brazil, paying for it myself. A scant ten days later, I received a call from a very happy Rachel who told me that the mockup fit perfectly in their aircraft and it looked really good.

About a week after this, John called me into his office to announce that he had read in SpeedNews, a well-known industry media fact-sheet, that Embraer had just closed a deal for their EMB-145 with American Eagle for over a hundred aircraft, and with Continental Express for about a hundred and seventy-five. He looked totally dejected.

"Too bad you no bid the HGS program with Embraer," he said.

"Well, I may have good news," I responded. "I forgot to tell Embraer that we wanted to no bid the project."

I also told him about the foam-core mockup, since I figured he would find out eventually anyway. He was definitely annoyed that I had "gone around him," but he soon got over it.

However, we still had the ugly NRE issue to find a resolution for—it was well over two and a half million dollars. I asked John for another trip to Embraer. He agreed to one more.

Back in Sao Jose dos Campos, Rachel took me immediately to their test aircraft, where they had managed to mount our HGS foam-core mockup to the cockpit. She was beaming—it looked amazing! She told me Satoshi was also pleased with it, and was anxious to discuss any NRE resolution I might have come up with while I was back in the US. I grew nervous.

Rachel took me back to Satoshi's office, where a friendly Beth led us in.

"Congratulations on your two new US customers for the aircraft," were my first words to Satoshi. He was obviously visibly pleased with the sale, but got right down to brass tacks.

"Did you come up with any way to solve our NRE problem, Phil?"

"With the modified Saab 2000 HGS, we are able to reduce the NRE to about two and a half million," I replied.

"That's still going to be a huge problem here," he continued, "and I don't know how to overcome it."

I suddenly had an idea!

"Satoshi," I said, "we both know that you can't amortize our NRE over fifteen or twenty aircraft, but what if we amortize it over three hundred?"

"I'm not sure I know what you mean," said Satoshi, obviously confused.

"What if we figure out an NRE payment for each new EMB-145 you deliver, until we get our NRE paid?"

"Even if they don't take the HGS?" asked Satoshi.

"Yes," I responded. I knew it would eat into their profit margin, but it could also gain a lot of new European sales for Embraer and keep their existing customers happy.

Satoshi was silent for a while. I could hear the gears grinding.

"It would be a much smaller number per aircraft," I pressed.

"Hmm, let me think about that idea." I could tell he was interested.

He told Rachel to take me to the Embraer cafeteria for lunch, and to come back mid-afternoon. I'm sure he had to discuss my harebrained idea with some other executives there. Over lunch, Rachel told me that she thought the idea might work, since Satoshi would have immediately shot it down if he thought otherwise. Meanwhile, I enjoyed my lunch with Rachel, and learning about what everyday life was like in Brazil.

Later we were back in Satoshi's office. He entered from a side door and shook my hand.

"OK, Phil," he said, "I really like your idea. We have a deal."

Uh-oh… I had blurted out my suggestion without first checking with headquarters to see if this idea would fly. I decided not to call John right away, but to wait until we were face-to-face in Oregon to allow him to reach across his desk and throttle me. I came up with all kinds of rebuttals on the long return flight to Oregon.

Back at our regular Monday marketing meeting, I updated everyone on the HGS mock-up fit at Embraer, and mentioned my NRE idea. John was a little puzzled until I explained exactly how it would work, and that we would get the NRE, but stretched out over three hundred aircraft deliveries. He was not as upset as I thought he might be. He said he would need to run it past our new owners to see how they felt about it, but in essence, he saw it as a potential solution.

By later that day, he confirmed that the new owners were OK with the deal.

With this one in the bag, it was time to turn our attention to new customers. We were off to the frozen turkey head races…again!

The Embraer ERJ family of regional jets – all with HGS as a factory option

America West B737-300 taxiing at Phoenix Airport. No fog!

A Regional Air EMB-145, the first HGS-equipped CATIIIa Embraer aircraft

Regional Airlines Embraer EMB-145 with our HGS installed

JetBlue Embraer ERJ –they became the first carrier to adopt dual HGS

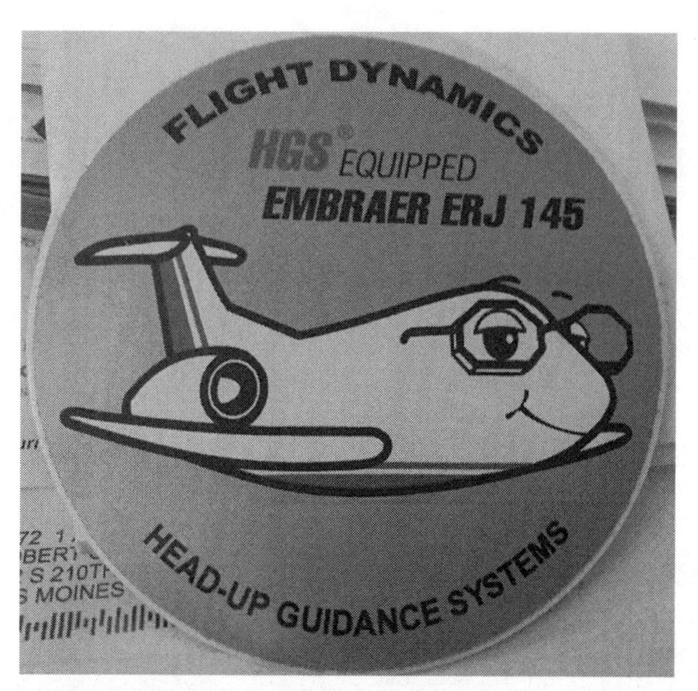

The Embraer EMB-145 fat li'l aircraft sticker

Flight Dynamics Christmas party 1997 (l to r) Jean and Dick Hansen, Chris and me, Sherri and Tom Geiger, Helene Bloch and husband

Chapter 17

Shades of Morris Air

After completing the Embraer deal, I once again felt on top of the world. There was lots of celebration at Barleycorns.

We had a great program manager at Flight Dynamics by the name of Bob Irish. Bob Wood assigned him to run with the EMB-145 HGS project, and to get the system JAA certified for Embraer's European customers. I knew the project was now in good hands.

In 1997, Kilbane and I were asked to attend the Paris Airshow, another big European aviation show. When we contacted Bombardier to set up some meetings with them at the show, they invited us to their big "bash" at the Eiffel Tower. We made sure to keep that evening free. John Howarth, who had been the original Horizon Air account manager for Bombardier, and his wife, Christine, were there, plus Rod Williams and Rod Sheridan, two great marketing folks who had been a huge assist to us from the early Dash 8 days, and Pete Freeland, one of our technical support folks.

They had also invited Joanna Speed from Speednews, the aviation news factsheet. Both Bombardier and Horizon hoped to get some good press coverage of CRJ and Dash 8 operators now flying CATIIIa approaches in Europe.

The whole evening was a blast.

I was still feeling good about the Embraer deal. I had arranged for Desmond and some of the Collins executives to meet Satoshi at the Embraer chalet at the Paris Airshow, and they all got along well. Satoshi was complimentary about my "innovative and creative thinking" on the NRE finance options, which I hoped would help offset any bad feelings John might still have about my rebellious behavior.

Many times after a good win, there was a bit of down time. But this time it did not last long. I was at my desk one day, relaxing and chatting away with Dick about the Paris Airshow, Brazil, and Embraer, when my phone rang. It was one of the original Morris Air Flight Ops folks, Skip Pennyweight. He had been a terrific assist during our campaign there.

"Phil, have you got a few minutes?" he asked me. "I need to update you on something."

"Of course, Skip. Okay if I put you on squawk box? I have Dick Hansen here with me and we're both interested in what you have to say."

"Sure," he responded. Skip knew Dick well. Skip had been Director of Training at Morris Air when we won the HGS program there, and when

Southwest acquired the airline in early '94. But he had decided not to join Southwest after a buddy offered him an attractive employment opportunity. He told us that he was now gainfully employed by Western Pacific Airlines, a startup based in Colorado Springs, Colorado, as their Senior Director of Training and Standards. He had been raving to the rest of his new team there about our HGS, and the success we'd had, not just with Morris, but also with Southwest and Alaska. He also told them what was happening at Boeing. They were very interested, he said.

"It would be well worth your time to pay us a visit here," Skip continued. "I can set it up for you if you like." Dick and I both agreed to go.

According to Skip, Western Pacific operated fourteen B737-300s and were in negotiations to lease or buy an additional six new B737NGs from Boeing. It wasn't a big fleet, but they were enthusiastic, and Dick and I felt that the trip to Colorado would be worthwhile. After Southwest's track record of success in the low-cost airline marketplace, you never knew where one of these small airlines might end up.

The company had originally been formed in 1994, and was now owned and managed by Ed Gaylord, of The Gaylord Entertainment Company. Their routes were mainly west of the Mississippi River, but had extended to the Eastern US and all the way to the West Coast as more B737 aircraft had been acquired. At one point, the airline even leased some B727 aircraft as well.

Western Pacific had no B737 autoland system upgrades, and they had some "sporty approaches" into Utah and their Colorado ski destinations. Skip had made a compelling argument for the new airline to consider our HGS's safety and operational benefits. He also felt that it would give them a leg up on their local competitor, Frontier Airlines. (During the campaign, we were asked many times if we were in discussions with Frontier about our HGS, and we continued to reassure them that we were not. Their relief made us wonder if we should be, but Frontier was rapidly becoming an Airbus operator, and we had our hands full with Boeing planes.)

Skip indicated that there was a fairly major problem that could really slow or even kill this early interest. They were a startup airline and were "cash poor."

"Could your company consider a leasing arrangement for the HGS on our existing fleet?" Skip inquired. Dick and I looked at each other in surprise. This was one question that had never been asked before in all our frozen turkey head efforts. But we felt that it should be explored, since there were bound to be others in the future with the same request.

A short while later, we were both flying to Colorado Springs. We met Skip in the lobby and he led us upstairs to their conference room and introduced us to the rest of his team. There was Tim Komberec, VP of Flight Ops, Randy Hodge, their financial analyst, Jerry Star, Chief Pilot, and Martin

Wax, VP of Purchasing. It was always a good sign to us when purchasing was invited.

We gave them HGS-101 and answered a lot of technical and business questions. Dick, as usual, added a lot of pilot credibility. Between the two of us, we were able to cover and solve every issue that came up. We told them that we were investigating their request about HGS leasing, but warned them that we had never done this before, and it would require the blessing of our company's senior management as well as our owners. We were unsure how it would be received.

We had two big benefits here. First, Morris Air had installed the HGS on their fleet of B737-300s themselves, and Western Pacific had hired some of the engineering and maintenance folks from Morris who had been involved in that install. The retrofit of Western Pacific's existing fleet would be based on the successful Morris Air model, which Skip and some of the other Western Pacific folks had a lot of familiarity with.

The second benefit was that, because other customers had paved the roadway at Boeing for factory install, Western Pacific's new aircraft, whether purchased or leased, could all be fitted with the HGS prior to delivery.

As for the leasing question, payback expert George Kanellis also had experience and expertise with customer aircraft leasing arrangements from his days at Boeing. We asked him if he thought Boeing could add our HGS to their leasing arrangement with the airline, but he convinced us that that route would take a very long time to be successful.

So Dick and I got together with George back in Oregon to discuss other leasing options. George very rapidly became key for this part of the project. Through some of his previous contacts, he managed to find a leasing company, Evergreen Financial, as a third party lessee, and began to work out all the contract details, something I wanted no part of.

In the meantime, Dick and I were back to our old tricks. Tim Komberec became even more of an HGS fan after a great M-Cab session at Boeing, and he and Dick got along really well. Tim was also pleased with their Boeing customer engineer's enthusiasm for the HGS. The engineer reaffirmed that Boeing would support any request for the system on Western Pacific's new aircraft, whether purchased or leased. Our earlier efforts with Customer Engineering at Boeing were really paying off.

Within a relatively short time, and after a few more simulator sessions for other key flight ops folks, Tim called to inform us that his team was now fully supportive, and they had made plans to present the HGS project to their CEO, Ed Beauvais, as soon as George had the final OK on the lease deal.

We upped the pressure on George.

A couple of weeks later, after approval from John and Collins, George had completed the HGS leasing arrangement. I forwarded it to Martin and

Tim, and explained that any new aircraft from Boeing could be added once all the final details, including quantity, were ironed out. After they had reviewed the arrangement, the Western Pacific team decided to proceed with their meeting with the CEO. This was great news!

We knew the date of the big meeting. That whole day, Dick, George, and I held our breath. Near the end of the day, Tim finally called to let us know that the CEO had approved the deal, including the new aircraft from Boeing! It was back to our favorite watering hole for another round of celebration. We took it as a good sign that these turkey head campaigns were getting a little easier—and more frequent!

The leasing arrangement was a first for Flight Dynamics, and we believed it would open up new doors at other startup airlines or cash-poor operators. I made sure that all of our Boeing customer engineering contacts knew about this leasing capability of ours, just in case it came up in discussions with other airlines.

For our win celebration with the airline, Tim found a very unique restaurant in Colorado Springs that served wild game. John, Dick, and I flew back to join the team for a great celebration dinner. There were lots of good jokes about what animals we were eating.

A short time later, Martin called to alert me to the fact that the Western Pacific regional feeder called Mountain Air Express (MAX) had acquired twelve white-tail DO328s from Dornier, and had options on twelve more (presumably from the contract Horizon had cancelled after the frozen wheel well doors incident). The aircraft were already provisioned for the HGS, and we began discussions about adding the MAX planes to our B737 lease deal. It looked like another winner.

Unfortunately, the Western Pacific project never took off for us.

About ten months later, in the spring of 1998, we were told that Frontier Airlines was in discussions regarding the potential acquisition of Western Pacific. This was a worrisome development. After my past experiences with Western Airlines and Northwest, I had become quite gun-shy about these acquisition deals. Most often, it meant the end of our HGS program.

After many industry rumors over the next little while about what was really happening, we read that Western Pacific was preparing to declare bankruptcy.

And they did! It was another terrible disappointment.

A Western Pacific B737-300 taxiing in at Colorado Springs

Western Pacific MAX DO328 taxiing at Colorado Springs

Tom Kilbane, Dick Smith, Ray Hillman, Helene Bloch, Chris, me, and George Kanellis, 1997

John Howarth, Pete Freeland, Rod Williams, Joanna Speed, Tom Kilbane, Rod Sheridan, me, Chris, and Christine Howarth at the Paris Airshow Eiffel Tower dinner

John Desmond (Flight Dynamics), his wife Jennifer, Chris, and me at the Paris Airshow 1997

Chapter 18

Getting Europe Going

Ever since he'd joined our team, Tom Kilbane had been a busy guy. Just a few months after he came on board, Southwest had made the decision to equip their entire fleet of B737s with HGS, which led to Boeing offering the HGS from the factory. "This is it!" Tom had exclaimed over a few celebratory beers. "We're on easy street now, and the orders from the airlines will come rolling in."

I reminded Tom that although I'd had similar thoughts many times in the past, the aircraft manufacturers, regulatory authorities, and airline finance people continued to make the HGS sales process quite challenging. It was still an uphill battle. But we all agreed that having the system offered from the Boeing factory was certainly a giant step in the right direction, and there was no question that airline customers would be enamored with Boeing's endorsement.

As the youngest person on our company's marketing team, and the only one who was still single, Kilbane had been tapped by John to focus on the European airline marketplace, which would require substantial travel and lots of time away from home. We all felt there would be great opportunities with the airlines there, especially after Dick, Jim, and I had travelled to Europe with the Boeing teams many years earlier to investigate potential interest in our HGS and had received such positive feedback. So that part of the world really needed dedicated attention. And we still had our list of contacts and interested European airlines. Tom was our man.

He had spent a significant part of 1995 working with Alitalia in Rome, Italy, on a possible HGS retrofit program for their MD80 fleet of around seventy-five aircraft. A nice-size order if he could pull it off. The HGS installation would be part of a major avionics upgrade retrofit that Alitalia was undertaking in partnership with the OEM, McDonnell Douglas, based in Long Beach, California.

After spending many months with the Alitalia flight ops team, educating them on all the benefits of our system and answering lots of questions, things finally boiled down to a decision between our HGS and the Sextant-Avionique HUD system. Sextant had been claiming that they had previously certified a hybrid approach—the combination of a CATIIIa HUD with a CATIIIa autoland system to achieve CATIIIb—on the Aeropostale B737-300. They were now offering this same approach for the Alitalia MD80.

Instead of an inertial reference system, or IRS, the French HUD supplier was using the existing AHRS, or Attitude & Heading Reference System, already installed onboard all the MDs. This system provided inertial data to the aircraft's cockpit displays that was similar to the IRS. Since the Sextant system didn't require the installation of an expensive IRS, it made the total cost of their installation substantially less than ours.

Our A-Team had looked hard into the AHRS as a cost-saving solution during Boeing's pre-implementation project in the late '80s, but Boeing had been adamant that it would not meet the strict FAA CATIIIa certification requirements. Alaska Airlines had also asked about using the AHRS for a large fleet of MD80s they were operating at that time, but we could not achieve the required accuracy for a CATIIIa landing system without the addition of an IRS.

In the end, the added cost of the IRS was too great, and Alitalia decided to go with the Sextant-Avionique HUD solution. I put on my Captain Sunshine hat and reminded Tom of some of my past fiascos and losses and encouraged him to "truck on."

While it was a crushing blow for Tom, and for Flight Dynamics, one good thing that came out of the Alitalia campaign was Tom's excellent relationship with Capt. Giovanni Riparbelli, the airline's Chief Technical Pilot. He was not enamored with the French HUD, and became a huge supporter of the HGS operational and safety features and benefits. He was also a terrific pilot, and was well-connected and well-respected in the airline industry. Over the following years, after his retirement from Alitalia, he was hired frequently by our company to help with HGS certification and customer training efforts around Europe.

About this time, I heard from one of my Boeing customer engineering buddies that Scandinavian Airline System, or SAS, was in early discussions regarding an order for forty B737NG-600s, with options for an additional thirty-five aircraft. They were interested in upgrading the aircraft's autoland system to CATIIIb.

At one time Boeing had looked into the autoland upgrade for British Airways. BA had rejected the one-point-five million dollar additional cost per aircraft, and Boeing was reluctant to go through the process again—the effort was huge, and Boeing was convinced that SAS would not like the final answer any more than BA had. HGS could be the solution.

SAS interest in the HGS was not new. My good BBJ friend, Borge Boeskov, had been Director of Sales for Scandinavia at Boeing when I had first been introduced to him by Chris Longridge. Being the HGS fan that he was, he had promoted our system whenever and wherever he could. So some initial groundwork had been laid at SAS.

By this time, Tom had made a convincing argument to John and our new owners that an apartment in Europe would be far less expensive than all the transatlantic airfares and hotels. So, with approval, he established Hoofddoorp, a town in walking distance to downtown Amsterdam, as our "European Headquarters" because it was home to the Collins European offices. He found a very cool apartment on one of the downtown canals, called Herengracht. My wife Chris and I crashed there one time while we were in Amsterdam on vacation and Tom was in Sweden. What a great location!

Jan Timmers, who was a big supporter of the HGS program, managed the Hoofdoorp office for Collins, and was kind enough to offer Tom some office space. Jan and Tom became good friends.

Jan came into Tom's office one morning and excitedly confirmed what we had heard from the Boeing customer engineer about some new B737NG aircraft for SAS. He added that the airline was now more interested in our HGS than in the expensive CATIIIb autoland.

"Hey, Tom," he almost shouted, "I do believe SAS may want to put your HGS on their new B737s!"

Tom wasn't sure if Jan was just exercising his well-known Dutch sense of humor, but when Jan gave Tom the contact info at SAS, and told him to call right away, Tom realized that Jan was indeed serious. The contact's name was Bengt Nilsson.

Bengt was the SAS Avionics Engineering Manager in Stockholm. He had reached out to Jan, whom he knew well, to ask about being connected to someone from Flight Dynamics for a chat about our HGS.

Tom was truly excited at the thought of working with one of Europe's premier airlines, and back at the plant we felt that, with their stellar reputation, SAS would be an ideal first major European airline HGS customer. Desmond told Tom to pull out all the stops. I shared my now-well-polished HGS-101 presentation with him.

On one of Tom's subsequent visits to Stockholm, Bengt introduced him to the SAS team that was working with Boeing to configure their new B737-600s and investigate customer options. There was Captain Jan Strom, VP Flight Ops, Ulf Johnfors, Director Flight Standards, Dick Jacobsen, from Fleet Planning, and Erik Juhl Andersen, Director of Purchasing. Bengt had all the right people involved.

While everyone knew that the B737NG had a CATIIIa autoland system, SAS was particularly interested in our HGS's reduced minima at CATI airports—there were many of those in Scandinavia—as well as the lower 300 RVR takeoff.

Around this time, Jan Strom was spending a lot of time with McDonnell Douglas in Long Beach, California, doing acceptance flight tests for the airline's MD90 fleet. On one of Jan's trips to Southern California, Tom invited him up to Seattle for an M-Cab session. With Dick Hansen's skillful demo approach and USAF background, and Jan's previous HUD experience from his fighter pilot days in the Air Force, Tom soon had a real HGS champion within SAS.

Before long, there was a full-fledged effort underway in Stockholm to analyze the costs and benefits of the HGS installation on the new SAS airplanes. Key to this effort was George Kanellis, who had done such a good job at Southwest. His HGS Economic Benefits Analysis (EBA) model had been even further refined, and would allow Tom to make a clear business case to SAS's senior management. George began to collect European weather and SAS flight schedule data to support Tom's effort.

Helping with the HGS business case from inside was Dick Jacobsen, from SAS's Fleet Planning Department. Jacobsen had been born in the US—Alabama no less.

"You can hear the southern twang in his Swedish accent when he speaks English," Tom laughed one day at our favorite watering hole near our facility in Tualatin.

These watering hole get-togethers were always a good opportunity for the whole marketing team to share our campaign progress and strategies, as well as funny stories. It also provided the opportunity to ask for advice or discuss solutions to issues that might have come up during other sales campaigns.

Dick Jacobsen had one objective—to make sure SAS didn't make any really big mistakes on SAS's new aircraft order. Tom went to great lengths to assure him that installing our HGS on their B737NG fleet would not be a mistake. We now had many industry references, and a number of B737 customers were routinely landing in CATIIIa conditions and were very pleased with our system's performance. We even had regional carriers flying to CATIIIa minima on Bombardier CRJs and Dash 8s in Europe, and they were all satisfied.

During their time in Stockholm, George and Tom enjoyed Dick Jacobsen's company immensely. They were thrilled when he invited them out to his summer home on the Swedish archipelago while they were working on the SAS economic benefits analysis effort over a weekend. They said it was a blast.

Dick had crunched the numbers and concluded that the payback period for the HGS investment satisfied the internal SAS criteria. He told Tom that he would recommend that the airline move forward with the project. This was huge. Getting Flight Ops folks excited about the HGS was one thing, but getting the "bean counter" from the Fleet Planning group to endorse it was another thing altogether—it was a major campaign accomplishment. I told Tom that Dick Jacobsen reminded me of John Owen, the CFO at Southwest, and his interest in our system—he had a lot of influence on the powers that be and the final positive airline decision.

In order for SAS to request Boeing installation of our HGS on their new planes, Tom needed Kurt Kuhne, SAS Director of Fleet Planning, and his team to make the recommendation to their Board of Directors for final program approval. After months of work, Tom was told he could obtain an audience with Kurt.

A key part of the presentation to the Fleet Planning group would be our economic benefits analysis, and to do justice to it, Tom felt he needed George Kanellis to be there. George had committed to making a presentation at a large aviation industry event in LA the day before the scheduled meeting with Kuhne. Being the team player that he was, and with some additional pleading from Tom, George said "no worries," he could get an overnight flight from LA and land in Stockholm in time for the critical SAS meeting. After his ten-hour-plus flight, George managed to shave and put on his suit before getting off the airplane. When Tom met him at the curb outside Stockholm's Arlanda airport, he was ready to go.

They raced to the SAS headquarters building in Frösundavik, just outside the city, and were quickly escorted into the conference room where Kurt and his team were ready and waiting.

This was 1996, in the days before fancy PowerPoint presentations with a laptop and a projector or an LCD screen. The presentation was printed on transparencies that were placed on an overhead projector. George and Tom were "in the zone" that day. The SAS team joked that they must have been briefed on the questions beforehand because each time someone asked a question, the next slide in the stack of transparencies contained the detailed answer. The meeting could not have gone any better. Tom and George were ecstatic, and the SAS team was extremely pleased.

GEC Marconi, who had done a real job on us at American Airlines, showed up at SAS in an effort to head off Tom's progress there. But Tom's established relationships with the Scandinavian team, and his efforts to convince them, won the day and prevented GEC from gaining any traction.

We received the endorsement of the Fleet Planning group, and the recommendation to instruct Boeing to install the HGS in the new fleet was approved by the SAS Board of Directors. The excited and happy SAS team graciously hosted a little ceremony to celebrate the decision, with a champagne toast for Tom and George.

Another milestone—our first major European airline customer!

Our company marketing team held our own celebration later at our favorite watering hole, although we preferred beer to champagne. Tom and George had done an outstanding job opening the major European airline market for us, and we toasted their success over a beer or two, or three.

One of the best things to come out of the SAS campaign, aside from landing a marquee customer across the pond, was our introduction to Capt. Jan Strom. Strom was a remarkable man and became a good friend to Tom and George. He had been a Swedish Air Force fighter pilot who'd been invited to attend the US Navy "Top Gun" school at Miramar as a young officer. He had also been a test pilot for SAAB aircraft, and Chief Technical Pilot for the SAS MD80 and MD90 fleets. He was an accomplished sailor, a scratch golfer, and a dedicated family man.

Jan passed away a few years ago, well before his time. Tom and George say it was a great pleasure to have known him and an honor to call him their friend.

For Immediate Release

Contact: **Helene Bloch, Flight Dynamics, (503) 684-5384**
Heleneb@fltdyn.com

June 25, 1996 (Portland, Oregon): Flight Dynamics Head-Up Guidance System Enters Service in Europe.

Flight Dynamics announces today the beginning of Head-Up Guidance System (HGS®) operations in Europe. The HGS has been installed in the Canadair Regional Jet, which was approved by the JAA in March 1996, and is operational with four of the five Europeans Regional Jet operators: Brit Air, Lauda Air, Lufthansa Cityline and Tyrolean Airways.

The HGS will allow lower visibility minimums for approach and landing operations by the Canadair Regional Jet. The 50-seat twinjet will be able to land in visibility conditions as low as 700ft RVR at airports that allow Category II or Category III approaches (200 m RVR).

According to John Desmond, Flight Dynamics President, "This milestone on the Canadair Regional Jet program confirms the growing acceptance of the Flight Dynamics HGS on commercial aircraft. The reduced visibility requirements at Cat II and Cat III airports enhances the appeal of the aircraft for these European airlines and provides them with the capability to stick to their schedules in low visibility weather. Arriving on time protects schedule integrity, keeps airline customers happy and increases airline profitability."

The SAS team with George and Tom, celebrating the B737-600 HGS win with a champagne toast

One of the new B737NGs being delivered to SAS with our factory-installed HGS

Chapter 19

China and Korea

After the acquisition of Flight Dynamics by Collins and Kaiser, we had many joint sales get-togethers with Collins to explore potential new airline opportunities and to educate some of the Collins team on the benefits of our HGS. In the process, we met a lot of their in-house and regional sales and marketing types, including Martin Lin, their terrific marketing manager for China.

One day in late 1997, Martin called John to alert us to a large Chinese airline jaunt. Collins would be presenting their latest products and technology to the airline industry in the People's Republic of China and Taiwan. Flight Dynamics was invited to go along and present our HGS. George Kanellis had recently been tasked with looking after China for us but was unavailable to go, so John tapped me.

Preparing for the trip required a little coordination with Martin Lin, who offered to interpret for me into Mandarin. This would require my speaking no more than two or three sentences at a time, and he would then translate. We went over my HGS briefing a number of times, and I explained what I was going to say for each slide. I wanted to be sure Martin fully understood our product and technology, and the intended message on each slide, to increase the chances of a correct translation and good audience understanding. Martin actually came up with the word for HGS in Mandarin, since the direct translation did not work in that language. I think the name he coined for it is still in use today.

Finally we departed for Shanghai, China, where our first meeting with local airlines would occur. It was to be a two-day affair. There were eight or ten airlines in attendance on that first day, and roughly five or six folks from each, so it was a good audience of executives, plus flight ops, training, standards and engineering, and maintenance personnel from the country's major airlines. I watched carefully that morning as the Collins folks went through their briefings with a translator, to learn how I would need to do mine. Finally, on the afternoon of day one, it was my turn. I got up and went through my slides slowly, providing time for Martin to do the translation. This stretched my forty-five minute HGS-101 into about an hour and a half.

There was mounting interest in the aviation industry at that time in the Enhanced Vision System, or EVS, technology—the superposition of an infrared image over a HUD's symbology. And there were lots of articles now appearing in the trade journals, mostly about progress being made by

Gulfstream, who intended to offer the technology on their business jets. Gulfstream's objective was to convince the FAA that the combination of HUD and EVS would allow for lower landing minima without a CATIIIa training and maintenance burden on the operator.

I had included a slide in my China presentation that showed our HGS symbology overlaid on an infrared image. It was actually quite impressive. As I went through my slides there was relative quiet in the audience until I hit the EVS slide. Suddenly, there was an absolute eruption of whispering, commenting, pointing, and discussions in Mandarin.

I stopped. "What the hell's going on?" I whispered to Martin. "What happened? What are they talking about? Why are they so excited?"

"I have no idea," he commented, as he tried to hear what they were saying. Neither of us could figure out what had just occurred. We decided to let it go and, as things quieted down, I continued with the HGS briefing.

Following our presentations that day, Collins had arranged for a reception and a delicious dinner for all of us and our airline guests. I managed to sit beside the Chief Pilot for China Southwest, which was based in Cheng-du. I cannot remember his name, but he spoke perfect and fluent English, so we had a great discussion about their airline, our company, the HGS, our customers, and Boeing.

Finally, when our discussion slowed down, I asked him about the outburst during my pitch.

"When I came to my EVS slide," I said, "why was there so much discussion and commenting going on in the audience?"

"Oh, that's easy," he laughed. "You see, to a Chinese pilot, a picture is worth a million electrons."

I never forgot that!

But the message was clear. He explained further that cultures around the world that were not brought up on our English alphanumeric system had to go through a translation process to fully understand the messages and numbers presented on the aircraft cockpit displays. If 10,000 was displayed for altitude, OK, but if it was for vertical speed, not so good! This translation process added reaction time which, even if only slight, could be a concern in an emergency situation. But when I showed the EVS picture, there was no translation required.

This message about the importance of images would stick with me, and would help me as I got more involved in explaining EVS benefits for the justification of our HGS.

We received much the same reaction at our session on the second day for still more airlines, and some regulatory folks from China's Civil Aviation Administration (CAAC).

We then flew on to Taipei, Taiwan, to present to a number of airlines operating there, including some freight carriers. All seemed similarly enamored with the EVS slide.

Back in Oregon, I made sure that our company management understood the impact that the addition of the EVS to our HGS had on that part of the world. Martin was doing the same up at Collins headquarters in Cedar Rapids.

I had made many good connections and friends at Boeing over the course of our efforts there. One was an engineer named Pat Harper, a great guy. He and I got along well and had many interesting chats in meetings and over lunches and beer. So when he called me at my desk one day in early 1998, not long after my return from China, I was all ears. He mentioned that Boeing had received a serious inquiry about our HGS from Korean Airlines in Seoul. He had been asked by his boss to attend a meeting in South Korea and was requesting that I accompany him. I brought it up with Desmond and he agreed.

Korean had about thirty new B737NGs on order, plus some options, and they had been asking Boeing all the right questions about our system. After coordinating with Pat and with our Collins rep in Korea, I boarded an aircraft for Seoul. I had arranged to meet Pat and the rep at the hotel downtown after our arrival.

The team gathered that evening at the hotel's bar: Pat, the Boeing and Collins in-country reps, and the Boeing customer engineer and sales executive who looked after the Korean airline. We strategized about the proposed agenda and the best approach. I was given the task of providing the HGS portion of the presentation for the meeting the next day, and was assured that the airline folks at Korean understood English well and there was no need for a translator. I intended to include the EVS slide that had generated so much interest in China.

It was a long day. There were initially seven or eight Korean Airlines flight ops and engineering personnel in the meeting, and another two or three dropped in later. The Boeing customer engineer did all the introductions to our team, explaining our positions and what organizations we represented.

Finally, it was my turn. I took the group through the HGS presentation which, by now, was very polished. There was much nodding in the affirmative while I spoke, but immediately afterward, there were so many questions that I realized not all of them had understood everything I'd said. The questions and discussions went on for two or three hours, and while most were technical in nature, and directed at me, many were for Boeing, regarding the price of provisioning and the actual aircraft delivery schedule. Others were for the Collins rep, about things like local customer support for the HGS and relevant test equipment.

Following the meeting, we invited our Korean hosts to dinner. It was a great restaurant recommended by the Collins rep, and we thoroughly enjoyed a delicious Korean barbecue meal—although I wasn't much interested in the kimchi. Pat wasn't big on it either, and we had a good laugh about that.

After the meal, some of our guests told us that it was customary in Seoul to go for some karaoke following dinner.

Sure, we agreed, why not.

But things got a little weird.

We were led to a building not far from the restaurant, with lots of happy banter along the way. Once inside the main entrance, we went down some stairs and through a long, dark, carpeted hallway into a large, well-appointed room. A huge crescent-shaped couch faced a small stage with a microphone and a screen—presumably where the words of the chosen songs would be displayed.

Immediately after we entered the room, some scantily clad women appeared and began ordering drinks. It was obvious they had been "assigned" to us when they came over and sat down right next to their "targets" and began to chat us up. Pat looked over at me and I could tell he was becoming concerned about possible expectations, as was I.

Shortly after the second round of drinks arrived, I got up and told the group that I was extremely tired after the long day we'd had, and felt I needed to get some sleep. Pat immediately jumped up and agreed with me. The rest of the group began to stand up to leave, but we reassured them that they could stay, as we didn't want to disrupt their evening. I told the Collins rep I would get the drinks. Holy crap, the bill was for over fifteen hundred dollars after just twenty or thirty minutes! I guessed that we were being charged for the girls' presence as well as their drinks. This one would be hard to justify back home.

Outside the karaoke bar, a very relieved Pat thanked me profusely for getting us out of there. Laughing a lot, we hailed a taxi back to our hotel. We would remember and chuckle about this event long afterward. We never asked, and never heard, about the outcome for any of the others.

Back at our office in Portland, I had some explaining to do about the fifteen-hundred dollar drinking bill. But John, having had a little experience with that part of the world himself, fully understood, and brushed it off with a hearty laugh. Whew!

A short time later, Pat called to tell me that Korean had signed up with Boeing for our HGS on their fleet of new B737s.

Our first Asian customer! We were now changing aviation all over the world.

HGS symbology superimposed over an infrared-based Enhanced Vision System (EVS) image

A new Korean Airlines B737NG during delivery

Enhanced Head-up Symbology Builds Situational Awareness

PAUL PROCTOR/SEATTLE

Flight Dynamics is testing performance enhancements to its Head-up Guidance System (HGS) aimed at improving pilot situational awareness, including rapid recognition and recovery from unusual attitudes or inflight upsets.

The new capability also addresses tailstrike challenges and offers precision takeoff and landing roll guidance in low-visibility conditions. The updates are scheduled to be certified in early 2000 as part of the new HGS-4000 system. Most of the improvements will be retrofittable to earlier Flight Dynamics' HGS systems (*AW&ST* Dec. 1, 1997, p. 54; Dec. 12/19, 1994, p. 50).

Some of the upgrades were requested by airlines while others are evolutionary, according to Flight Dynamics President John Desmond. Alaska Airlines, Delta Air Lines and Scandinavian Airlines System formed an HGS advisory group about eight months ago and have been working closely with the Portland, Ore.-based head-up display manufacturer, he said.

Alaska uses the HGS on most of its 45-aircraft 737 fleet and is a launch customer for the next-generation Boeing 737-900, which at more than 138 ft. in length is the longest 737 built. Delta has ordered HGS for its new 737-800 fleet and has options to install it on its growing 777 fleet and in stretched 767-400s it has on order. SAS is equipping its new 737-600s with HGS.

European airlines flying the 50-seat Canadair Regional Jet also have been big purchasers of HGS. It allows them to operate more reliably in the bad weather that frequently plagues many airports in Europe.

This *Aviation Week & Space Technology* editor evaluated the latest HGS enhancements during a multiple-approach, 3-hr. session in Boeing's full-motion M-Cab development simulator here recently.

Tailstrike display on landing alerts pilots that they are in jeopardy of a tailstrike during strong, gusty wind conditions or when encountering wake turbulence from operations on a parallel runway.

Already familiar with the HGS-2000, I found the new functions seamlessly integrated with Flight Dynamics' existing monochrome-green HGS symbology. They were effective and easy to use, and I made the transition to the upgraded system quickly.

AFTER A SHORT "DIFFERENCES" briefing, I entered the simulator with Desmond and Dick Hansen, who serves as director of flight operations for Flight Dynamics. I took the captain's seat, which was HGS-equipped, while Hansen took the right seat and worked without an HGS and only a small symbology repeater display.

Airlines initially were interested in HGS for use in low-visibility conditions, including qualifying to operate at lower weather minimums. But now they are increasingly recognizing HGS' value for improved pilot situational awareness in all phases of flight, Hansen said.

Overall, Flight Dynamics has about 1,000 HGS units on order from or in service with 15 airlines. A new dual HGS installation is flying on U.S. Air Force C-130Js and is part of the avionics modernization program planned for the service's existing C-130E/H fleet, Desmond said.

After a few minutes of hand-flying to reacquaint myself with the standard HGS display, we began a good-weather, hand-flown ILS approach to a generic airport.

During the approach, Hansen pointed out some of the new HGS symbology—a small analog-style angle-of-attack (AOA) circular gauge located at the two o'clock position of the display. An index mark shows the best AOA for an approach. In production versions, the index mark will be located at the standard three o'clock position, to help the pilot easily interpret the instrument during emergencies or periods of high workload, Hansen said.

A representation of AOA stickshaker activation point is at the top of the gauge's arc, and the digital readout indicating AOA is adjacent to the gauge. It was not properly calibrated for our flights, however, due to simulator configuration limits.

Having the AOA gauge on the HGS display during approach gives pilots an added cross-check against mis-selected approach data, airspeeds or aircraft estimated weight entered into the flight director, Hansen said. As with a head-down AOA display, it also provides vital information needed by the pilot to extract maximum performance from the aircraft or in event of an air data system failure.

Upon touchdown, unnecessary flight- and navigation-related data disappeared from the HGS display, leaving mainly the horizon line and side-located airspeed and altitude tapes optimized for the rollout maneuver. The HGS flight path cue also changed from a bent-wing to a straight-wing representation, denoting on-ground operations.

Simplified, straightforward unusual attitude display "sphere" provides the information required to return the aircraft to controlled flight. Display intentionally resembles an Attitude Directional Indicator.

Hansen didn't need to point out the new symbology here. In concert with the raw data localizer line, a small circular rollout guidance cue appeared in the center of the display.

To correct drift and bring the aircraft back to the runway centerline, the pilot needs only to "step on the ball," or depress the rudder pedal on whichever side the guidance command circle drifts from center screen. The object is to move the rollout guidance cue into the center of the larger, "straight-winged" flight path symbol, now functioning as a ground-roll symbol. The technique is an instinctive one, similar to the centering of the turn-and-bank indicator ball.

Also upon touchdown, a digital callout appeared at the lower right center of the HGS display indicating runway length remaining. It clicks off distance in 500-ft. increments. The countdown starts—although may not be visible—once the aircraft passes the runway threshold, Hansen said. The data appears only after touchdown and is a computed value based on runway length input to the HGS control panel during preparation for the landing.

On a second, similar approach, I was able to more smoothly transition from short final through touchdown and rollout, with the correct flare point cued by the HGS' standard crossed-line symbol. Although just introduced to the rollout guidance capability, I found that assimilation and responsiveness to the corresponding symbology was easy.

A third approach and touchdown demonstrated the value of the rollout guidance feature in extremely low visibility. The Category 3a landing conditions we programmed into the simulator dramatically showed how easy it is to lose runway situational awareness during the rollout maneuver. A subsequent approach, in better visibility, but with an 18-kt. direct crosswind, helped illustrate system robustness. After touchdown, the rollout guidance cue helped me compensate with the rudder pedals as the stiff crosswind tried to push the aircraft off centerline.

WE THEN BEGAN A SERIES of takeoffs, demonstrating HGS takeoff-roll guidance in very low visibility conditions. It functions similarly to the landing rollout feature, providing command guidance to track the centerline. I also found the runway-remaining data useful to monitor aircraft takeoff performance.

Another piece of new HGS symbology came in handy on the subsequent takeoff attempt, performed under 300-ft. runway visual range (RVR) conditions. Hansen demonstrated a rejected takeoff (RTO), with the decision made at about a 120-kt. ground speed.

Upon initiating the high-energy RTO, a deceleration rate index appeared. It is a caret located just below the aircraft reference symbol, or boresight. As brake pressure is increased, the caret translates vertically downward past tick marks labeled -1, -2, -3, and finally to a "-MAX" symbol. The caret indicates braking effort and is roughly indexed to similar autobrake settings, Hansen said. This display allows more effective manual brake modulation by pilots during both normal and emergency stops, he added. It also allows the flightcrew to monitor the performance of autobrake systems, if selected. The deceleration rate is computed by the HGS using raw inertial reference unit data.

For the next takeoff, Hansen programmed an engine failure after V_1 or de-

Takeoff rollout guidance cueing is shown during a high-energy rejected takeoff. The guidance cue is nearly centered within the aircraft ground-roll symbol. The caret translates vertically to indicate braking effort.

Chapter 20

O Canada

Our Morris Air buddy and enthusiastic HGS fan, David Neeleman, after leaving the Southwest board, had gone to Calgary, Canada, in 1995 to establish a new low-cost carrier there, and called it WestJet. I had visited the airline once, sometime in late 1997, and had met some of their flight ops folks, including a Captain Stu McLean, their Manager of Flight Ops Engineering. Another nice guy! I gave him HGS-101 and answered his questions about our company and the HGS technology. He was especially keen on finding out more about what Boeing was up to with our system. Afterward, I asked him if we could go and see David, but was told that he had already left WestJet, and was starting up yet another new airline in New York, called JetBlue. But he had left his HGS fingerprints at WestJet.

In 1998, with me busy with Boeing, Korean, and the China visit, George Kanellis offered to continue the chase at WestJet. I told him I thought that the airline was ripe as an HGS customer, and provided him the background on my discussions there, along with Stu's contact info.

George began to visit them on a regular basis, and learned quite a bit more about the organization. Their original corporate objective was to become the "Southwest Airlines" of Canada. They had launched into the no-frills airline business and managed to meet or exceed all of their corporate goals in both growth and profitability. By late 1998, the airline had established itself as a solid low-cost carrier, and was well on its way to achieving success.

The airline's executives had managed to stay on track and had a full head of steam to take on new routes, most of which would be competing directly with the country's well-established and largest carrier, Air Canada. The country was receptive to alternate carriers, since most Canadians felt that Air Canada had a virtual monopoly and were taking advantage of it.

Stu introduced George to his boss, Bruce Flodstedt, their Director of Flight Operations, who was to become another strong advocate for our HGS.

WestJet's current fleet consisted of older B737 models, but they were in serious discussions with Boeing about a large order for newer aircraft to replace their aging fleet. By mid-year, they announced their intent to purchase twenty-nine new B737NGs. WestJet was growing, and George clearly saw the opportunity and began to plan his win strategy.

Stu's job at WestJet was to complete the B737NG configuration by selecting from the options offered by Boeing. After my HGS-101 introduction, he wanted to know more about the Southwest selection of our

HGS, and their pioneering efforts at Boeing. He had read some positive industry articles on our system, and he had been in touch with Southwest flight operations people in Dallas several times. It really helped Stu's case that new aircraft could now be delivered from Boeing with the HGS already installed.

Tom Kilbane and I shared many of our secrets of success with George during our watering hole get-togethers. Tom's most recent success at SAS, and mine with Korean, each offered some international customer strategies and ideas that we sincerely hoped would help George in Canada. I respectfully suggested that he stay away from any karaoke bars…

Stu introduced George to the airline's current president, Tim Morgan. Tim had followed in Neeleman's footsteps and, after discussing it with David, had become keenly interested in what the HGS had to offer their blossoming airline. George spent considerable time bringing Tim up to speed on what was happening at our company and, especially, at Boeing.

The good news was that the HGS could be included in the price of new airplanes, but the bad news was that, following the American Airlines selection of the GEC Marconi VGS, and Boeing's activities to accommodate that system for factory installation, the GEC HUD option was also in the Boeing catalog. But the selection of the HGS by Southwest had a huge impact on the "look-alike" WestJet and, thanks to a lot of hard work by George, they were definitely leaning in our direction.

In WestJet's route network at that time, weather disruptions usually referred to snow, ice, and blizzards. Fog was only an interruption in the farthest east and west provinces of the country. So most flight disruptions were related to the cold Canadian weather climate conditions and not to poor visibility. They were not yet flying to a lot of US destinations.

With a similar percentage of fog issues, Southwest had managed to justify the HGS because of their very tight operational schedules and short turn-times. Our CAT IIIa-capable HGS minimized the ripple effect of any poor visibility disruptions anywhere in their system.

WestJet believed that the extensive effort Southwest had gone through to evaluate the HGS was sufficient, and there was no need to plough the same ground again. After much consideration, and with George carefully explaining the benefits of our system and its advantages over the competition, WestJet finally decided on our HGS for their first twenty-nine new airplanes from Boeing. Stu called George to give him the good news, and they called Boeing to approve the installation of our system on the new aircraft delivered to Calgary.

We had our first Canadian airline HGS customer!

It was off to the watering hole again for some celebrating, and George entertained us with a funny story. During the WestJet campaign one day, after their meetings at the airline's facility, George and Stu went to the Calgary Stampede to see the "Chuckwagon Races," an event that is unique to this annual festival. (George had forgotten to get any Canadian money so Stu, ever the perfect host, ended up paying for their entrance to the event and everything else at the fair.) At the start gun, a team of horses is hitched, a field stove is loaded into the wagon, and the horses race off. A rather wild event, but it seemed to make sense to the experienced and excited Canadian attendees. WestJet actually sponsored one of the teams, and their team qualified for the finals. Everybody seemed happy.

At the sight of all the thick dust stirred up by the horses and wagons during the race, George leaned over to Stu.

"You know, if they were to put our HGS with EVS on those wagons, they'd see a whole hell of a lot better in that poor visibility," he chuckled.

"You're probably right," replied Stu, "I'll have to speak to them about your idea."

They both had a good laugh.

WestJet took delivery of their first new HGS-equipped airplane in May of 2001. George was invited to go along on the flight from Seattle's Boeing Field delivery center to Calgary. Stu and Tim Morgan performed the flight crew duties. About halfway to Calgary, Stu came on the aircraft speaker system to discuss the weather conditions.

"There are some broken clouds up ahead—but we expect to fix all of them prior to our arrival!"

Everyone got a good chuckle out of that. But Stu was right—the weather was perfect on arrival into Calgary. There, a large crowd of WestJet employees gathered in the hangar to take a look at their new aircraft, which also sported our HGS. Stu was keen to explain the system and its benefits to any and all interested employees. George was very pleased indeed.

But then, a big problem surfaced.

For some reason, the airline continually delayed implementation of the pilot training program on our system. This worried us a lot. Most HGS customers started the basic training as soon as the acquisition decision was made and a system was available in the simulator. WestJet's delays would result in little to no pilot familiarity with the system on their newly arriving fleet. It was particularly upsetting for George, who had spent so much time and effort on this project. We had even been able to solicit some training support from Southwest…all to no avail.

For the next twelve months or so, WestJet still could not be persuaded to initiate a thorough and professional pilot training program to support the HGS becoming operational. We could not figure it out. Every single HGS order up to that time had been met with excellent pilot reception, and flight ops training had never been extensive or burdensome to the customers.

Despite George's best efforts, WestJet's reluctance continued, jeopardizing the successful implementation of the system into daily line operations. It became obvious to George that there were two groups inside the airline, one pro-HGS, and another anti-HGS. Stu McLean was still our champion and tried his best to "convert" the anti-HGS folks, but he was essentially overruled. We were never able to learn the reasons behind the anti-HGS group's opposition. The airline had received the HGS on their new aircraft, but did no pilot training to make the system operational. It was unbelievable!

Sometime later, Bruce Flodstedt, who had now been promoted to VP of Operations, asked for a meeting with George to discuss returning our systems back to Flight Dynamics. We were shocked. This would require us to develop new engineering procedures to replace the HGS-modified flight deck panels, and to reconnect the flight deck interfaces with other equipment—not a trivial effort or expense. Since we had no other customers for those "used" units, and since we had no experience with the removal of a Boeing factory-installed HGS system, it was of no value to our company to support his request. We politely declined. Bruce jokingly intimated to George that perhaps WestJet could sell the systems on eBay. Ultimately, WestJet never used the HGS, and never removed them from their aircraft either.

Later, we learned that the aircraft were leased to WestJet through GECAS, a large aircraft leasing company based in Shannon, Ireland. WestJet was not even authorized to remove the HGS units. It was definitely one of our weirdest sales experiences.

The whole thing was frustrating not only for George and our A-Team, but particularly for Stu and some of his supportive colleagues inside WestJet. Canada had a significant number of CATI airports where our HGS lower landing minima would have been a godsend. Also, WestJet later opened up a lot of new destinations in the US, on both the east and west coasts, where the frequency of fog became a significant issue, and where the lower takeoff minima would have added significant value to their operations.

It was really a shame, and it remains unexplained to this day.

Stu McLean of WestJet and George Kanellis in the WestJet simulator

WestJet receives a Boeing B737NG with factory-installed HGS

Most WestJet weather disruptions were caused by tough Canadian winter conditions

Chapter 21

CHP

In early 1998, American Airlines selected the GEC Marconi Visual Guidance System (VGS) for the HUD option on their substantial order of new Boeing B737NG aircraft. American knew that after Southwest Airline's HGS purchase, Boeing was offering HGS provisions on any of their new B737NG aircraft delivered from the factory, and they knew that Delta and Alaska had also signed up for factory-installed HGS.

But American had selected the VGS, and for their huge order of about three hundred aircraft, they insisted that the Boeing factory install the VGS provisions on these planes.

In the past, I had gotten along very well with Bob George, who had done such a great job for us while he was the Southwest Customer Engineer at Boeing. When I contacted his Boeing office one day to catch up with him over lunch or a beer, they announced that he had left Boeing and was now working for AT&T. What the...?

Someone at Boeing gave me his new telephone number and I called him. I thought that with all the current activities at Boeing, and my added responsibilities with other airline customers, we could use someone to help us in Seattle. Bob already knew a lot about our system from working so closely with us for Southwest and at Boeing, and he jumped at the chance to join our team. I told him I would have to clear it with John Desmond first. John had already met Bob many times during the Southwest campaign and the subsequent LUV Classic barbecues, and he was receptive to an exploratory discussion with him. I gladly set it up, and Bob joined our growing and illustrious sales and marketing team in July of 1998.

From his station in Seattle, he kept us posted on the situation at Boeing. I was very glad that we had hired him, with his background at Boeing, his contacts, and his knowledge of how things worked there.

He called me from Seattle one day. "Hey, Phil, Boeing is really getting wrapped around the axle with two major airline customers demanding different HUD provisions on their new aircraft. It's going to require mammoth changes to their manufacturing processes."

"So...?" I was not sure what Bob was getting at.

"They're considering going to a common set of aircraft provisions for all HUD suppliers," he continued. "They want to do this for all aircraft in production at their facilities."

"Holy crap!" I responded. "That could require some major changes to our

HGS." (But I also knew that it could open the floodgates for selling our system on other aircraft types.)

"Yep! Just thought I should alert you so you can advise the A-Team that it's coming, and likely soon. Not sure what we can do to prepare, but they should be warned," he said.

I took his always-appreciated advice, and briefed the A-Team. They became more than a little concerned. From many points of view, this was a major complication.

Meanwhile, Wichita was screaming down the walls. With the VGS, they would now have two separate Section 41 HUD installation procedures to manage, and they were still working out the kinks on the first one.

In the initial HGS factory provisioning installations for Southwest and Alaska, there were some misalignment issues when Wichita tried to install the HGS hard points in the cockpit section prior to mating it to a fuselage. The hard point installation required perfect alignment of the HGS with the centerline of the fuselage, and installing the hard points in the cockpit prior to fuselage mating caused major problems with the alignment procedure. Two of Alaska's aircraft deliveries were actually delayed by the problem, which was anathema to Boeing.

Once the assembled fuselages arrived in Renton, there was the issue of different HUD system wiring for different customers, depending on what other optional avionics equipment they had selected. Boeing was going nuts! Having Bob in Seattle became a tremendous assist for our company in navigating these troubling issues, since he knew many of the prominent and relevant players, as well as the company's internal processes and procedures during airframe manufacture and assembly. He was able to keep us completely updated on what was going on.

Boeing's procurement management folks perceived Flight Dynamics as having a virtual monopoly on the commercial HUD marketplace, and thus on the pricing, and this was causing them grief. Lack of supplier competition was not the "Boeing way." So they encouraged Boeing engineering to support the American Airlines request for the GEC VGS in order to create some competition with our HGS, and get our prices reduced. I became concerned that offering the VGS on Boeing planes might also have an impact on our Boeing Business Jet program. BBJ was now offering other types of commercial air transports as bizjets, and if VGS were a factory option, there was a risk customers might choose that instead.

According to Bob, Boeing was also still not enthusiastic about supplier STCs. We knew that from earlier discussions with folks there, and from Alan Mulally's resistance to using them on Boeing's production line. Engineering groups believed that the FAA had a double standard for certification—a very strict one for the OEMs, and a much more lenient approach for suppliers. That was why Boeing often wanted to reengineer any STC brought to them by an airline or supplier. Safety was a big driver.

Boeing hoped that Common HUD Provisions would allow Wichita to return to the cookie-cutter method and avoid many of the logistical problems occurring during the HUD provisioning process, while accommodating airline requests for either system. CHP would also allow Boeing to prepare for future customers that might ask for a system on other aircraft, such as the B757 or B767. In fact, Delta had already asked Boeing about HGS on their new B767-400 order, the latest variant of that wide-body aircraft.

Alan Mulally had handed down an edict that said any HUD installed on one Boeing aircraft had to be installable on all of their aircraft. So CHP had to apply to all aircraft in production at Boeing's facilities, including the new B777.

Boeing began the detailed effort of developing their CHP specification. There was concern that Sextant-Avionique might get a HUD commitment on new aircraft from an airline in Europe, further complicating the whole process, so they had to be kept in the loop. However, since Sextant-Avionique did not have a current B737NG customer, the CHP effort focused mainly on Flight Dynamics and GEC. Boeing formed a CHP Advisory Group consisting of folks from both HUD companies. As any reader can imagine, that was really fun!

Boeing next issued a draft statement of work (SOW) that included mechanical and electrical design inputs from both Flight Dynamics and GEC. Preliminary CHP design began about a month later and was aimed mainly at the B737NG, since there were actual customers for this aircraft.

For starters, in order for the CHP guidelines to accommodate either supplier's hardware, both the Flight Dynamics HGS and the GEC Marconi VGS mounting schemes had to be modified to be the same.

There would also need to be wiring changes, both to the aircraft and, in all likelihood, to each supplier's HUD. Our A-Team was not happy about this, and I'm sure it was the same at GEC. At one meeting called by Boeing to discuss HGS changes, there was some loud rumbling around the table as our technical team indicated displeasure in having to "detune" our proven and industry-accepted installation engineering to accommodate a competitor's product.

The new HUD cockpit provisions also had to consider all the current aircraft structural design constraints, such as bird-strike, air vents, sun visor, map lights, and a myriad of other critical elements.

Then there was the flight crew interface—how the pilots were to interact with the HUD systems, and what controls were needed for mode changes, brightness control, blanking, and more. We had our own dedicated HGS control panel, but GEC were using a modified MCDU (Multi-function

Control & Display Unit) from their original partner, Honeywell.

The electrical interface required a very detailed investigation of which systems had to interface with the HUD, and what impact that might have on the system's other functions and performance. Then there were cooling requirements, EMI (Electro-Magnetic Interference) and HIRF (High Intensity Radiated Fields) impacts, and the need to develop a completely new wiring system to include warning lights, circuit breakers, and any general cockpit control devices. All of this was extremely complex, especially since there were five different Boeing aircraft in production at that time.

The CHP hard point provisions would not necessarily be the same for each airplane model, due to major airframe structural differences, but the intent was to minimize, as much as possible, the impact of variability on the airframe production line.

The pilot interface for each airplane model would also be slightly different, but the intent, again, was to use a set of overall system provisions that were essentially the same across the different aircraft types.

Because the specification for the CHP project was being created for Flight Dynamics and GEC Marconi on behalf of their customers, Boeing planned to charge these suppliers for any non-recurring engineering efforts to incorporate the configurations.

Boeing's terms for their NRE were initially a bit stiff, given that the CHP contained the OEM's commitment for every airframe in production. Neither Flight Dynamics nor GEC Marconi wanted to pay for CHP development on other airframes without a committed airline customer and confirmed product sale. So a payment schedule was worked out in which Boeing only got paid for the CHP for a new aircraft type once there was customer commitment to a supplier's HUD.

An interesting situation arose with the 777. Honeywell had scheduled a "blockpoint update" to their Aircraft Information Management System (AIMS), the main avionics computer onboard. This update would inhibit future aircraft HUD operations, unless the CHP partners ponied up to have Honeywell include certain required HUD changes to their software code. We pushed back mightily but, ultimately, we had to pay for the updates.

On top of all this, Boeing was asking for royalty payments on HUD provisions going forward. It looked like paying the NRE was only the beginning—sharing in our future profits was the other shoe dropping.

Finally, there was certification. Boeing had agreed to modify their Type Certificate for each aircraft type to include the HUD hard point provisioning, wiring, and other changes specified in the CHP. Many airlines would not accept a supplier's STC—they wanted the Boeing stamp of approval—so a certified Boeing Type Certificate would go a long way to opening some new doors for us.

The target first airplane to be equipped for CHP would be a B737-900 scheduled for delivery to Alaska in March, 2001.

The CHP effort revealed that our newest model, the HGS-4000, had far fewer issues than the GEC VGS in fitting into any of the Boeing aircraft (except the B747, which I'll get to in a minute). I was surprised at this outcome, since I had assumed that GEC's unique tray design would make their units more readily compatible with various other aircraft.

As for the B747, Boeing had supplied our A-Team with some CAD files of the B747 overhead structure, and we were shocked to find a huge I-beam member immediately above the flight crew's heads. The fuselage/cockpit cross-section for this aircraft was not round, as all the other aircraft were, because it had to accommodate the upper level flight deck. This resulted in more of an oval shape, and the I-beam was needed to accommodate irregular pressurization loading. The A-Team investigated potential HGS overhead unit designs but, in the end, could find no simple solution. It looked like the HUD symbology would have to be projected down and under the I-beam structure, likely to a mirror, and then back up to a prism of some sort to get the symbols back to the combiner. A real nightmare! Luckily, there were no obvious customers for the HGS on this aircraft.

Much of the final Boeing CHP development and success occurred after I decided to leave Flight Dynamics. But that decision is a story for the next chapter.

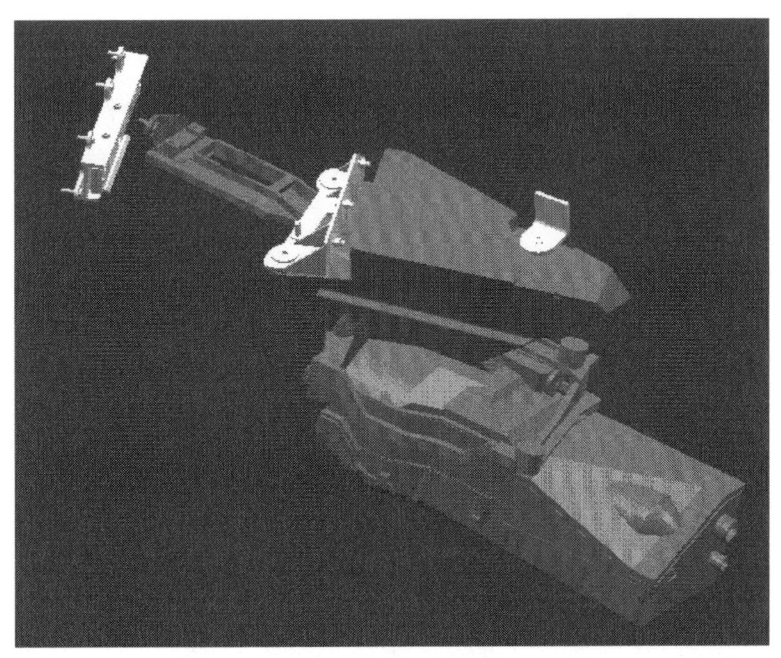

CHP had to accommodate both Flight Dynamics (top) and GEC (bottom) HUDs

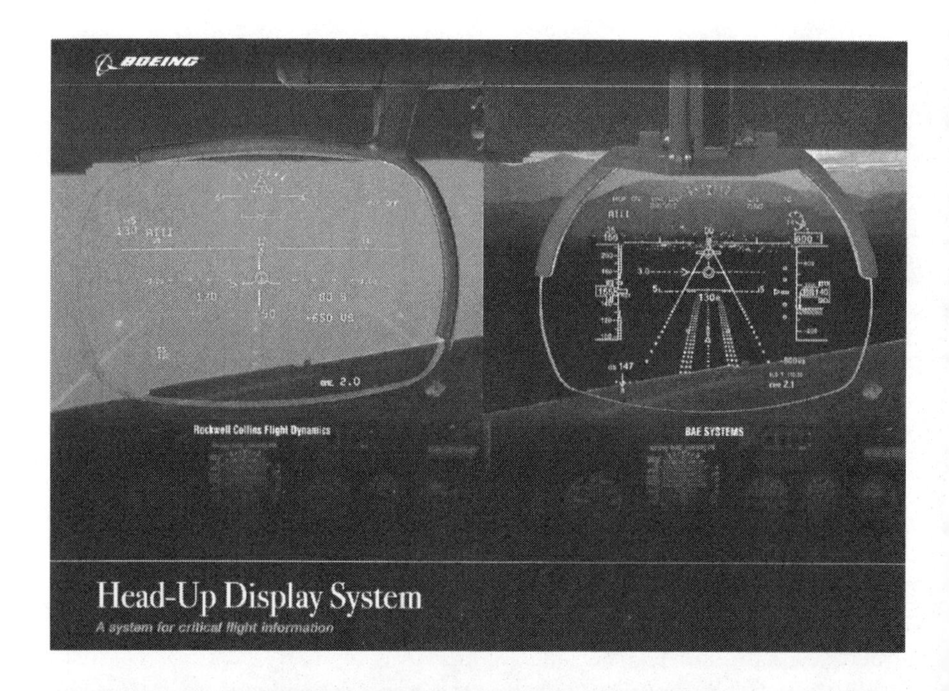

BOEING

Head-Up Display System

A system for critical flight information

Integrated with airplane systems and linked to the inertial reference unit, the Head-Up Display (HUD) collects and displays critical flight data. Airspeed, attitude, altitude, velocity vector, command guidance information, and flight path acceleration are translated into easy-to-read symbols that are projected onto a transparent glass combiner in front of the pilot. Some symbols, such as the horizon line and velocity vector, are conformal – they line up with the outside world. Pilots can simultaneously compare critical flight data with real world conditions.

Operational benefits include manual Cat IIIa landing capability and low-visibility takeoff. With HUD, airplanes can arrive and depart safely and on schedule, despite weather-related visibility problems that would otherwise disrupt operations. By minimizing delays and increasing passenger satisfaction, airline operators achieve significant economic benefits.

HUD enhances situational awareness and safety. By presenting flight data in the forward field of view, HUD allows pilots to look outside and simultaneously view critical flight parameters, flight guidance cues, and the airplane's flight path. HUD provides visibility of the airplane's energy state and speed trend, enhancing stabilized approach to land. HUD also provides improved manual touchdown precision, tailstrike avoidance, and unusual attitude recovery guidance.

HUD creates a platform for future innovations such as enhanced vision, synthetic vision, and taxi guidance. Enhanced vision will project both infrared and radar images on the HUD combiner and will display a view of the landing environment through clouds or fog. Synthetic vision will create a high-quality digital image of the surrounding terrain. Taxi guidance will provide improved surface operations during periods of poor visibility.

Boeing now offers airlines the option of installing HUD systems manufactured by either BAE SYSTEMS or Rockwell Collins Flight Dynamics. Installation is now available for the 737 family and will be offered sequentially on the 737, 777, 717, and 767. Boeing is studying the possibilities for the 747 family.

Rockwell Collins Flight Dynamics HGS4000

BAE SYSTEMS VGS 2020

Safety and Operational Benefits

- *Stabilized approach to land*
- *Unusual attitude recovery guidance*
- *Constant view of flight path, energy state, and speed trend*
- *Improved manual control for touchdown precision*
- *Manual Cat IIIa approach and landing capability*
- *Low-visibility takeoff guidance*
- *Precise speed and acceleration control*
- *Platform for future technology*
  - *Enhanced vision*
  - *Synthetic vision*
  - *Taxi guidance*

The Boeing Company
Commercial Airplanes
Marketing
P.O. Box 3707
Seattle, WA 98124-2207

www.boeing.com

Printed in U.S.A.

The Boeing CHP effort resulted in a brochure covering both HUDs

Chapter 22

Time to Move On

Nearing the end of the '90s, two things happened that had a significant influence on my future at Flight Dynamics.

The first occurred at yet another NBAA show, in 1998. We were allowed to exhibit our HGS in a corner of the Collins booths. I invited a couple of senior folks from Atlantic Southeast Airlines (ASA), the regional feeder for Delta, to come and see our HGS. I thought they could easily become a Flight Dynamics customer, and I was eager to gain another US regional operator after our Horizon success.

I had been to visit ASA at their facility in Atlanta a couple of times, to do the now-familiar HGS-101 and to answer questions from their group. As a Delta feeder, they had a situation similar to that of Horizon feeding Alaska Airlines, with the same logistical impacts during fog events. Our Delta buddies had done a great job convincing them to look seriously at the HGS. With the system now offered on the Bombardier CRJ from the factory—an aircraft operated by ASA—their senior management had become quite interested in our HGS. They would attend this year's NBAA and wanted to come and see our system.

The start of the show was a little disconcerting. Our owners went through a pitch about how we were all supposed to stand in the booth—no hands in pockets, no crossed arms—which we thought was a bit weird. They even gave us a printout showing the various unacceptable positions with a red line slashed through them. Strange!

While hanging out in the booth by our HGS, I saw Bryan Labreque, a Senior VP for ASA, and their president approaching. I had met both of them at their headquarters in Atlanta during my briefings there, and Bryan and I seemed to really connect. I made my way into the aisle to greet them.

"Hi folks. Welcome! Are you here to see our HGS?" I asked enthusiastically.

"You bet, Phil, but we need to meet with another supplier first...nothing to do with your product," Bryan said. "We'll come and find you very shortly. We're really looking forward to seeing your HGS and to catching you up on things at our place, so don't go too far away."

"OK, great, I'll be here," I replied, walking casually back to our corner.

When I returned to the booth, I was severely chastised in a loud voice by one of the robust salesmen Kilbane and I had encountered previously at the Farnborough airshow. I could tell immediately that he was not a full-fledged member of the “Captain Sunshine fan club.”

“Moylan,” he yelled at me gruffly, “we do not accost our customers like that at an airshow.”

“It’s OK, I know those folks very well. They’re here to see our HGS for their regional fleet, so I wasn’t accosting them,” I replied, as nicely as I could.

“I don’t care,” he continued very loudly, “it’s not OK, and that’s not how we operate here. You better get used to it.” He stood glaring at me for a good five minutes after the episode.

I became very worried that I could never fit in with this mentality. Later at the show I spoke to Desmond and told him that I did not think my customer approach was going to fit in well with our new owner’s sales philosophy. I told him I was becoming quite alarmed, but he reassured me that all was OK. Still, it stuck with me.

The second thing that would influence my future at Flight Dynamics was a rumor that Collins was going to buy out Kaiser, our other owner, and would own a hundred percent of Flight Dynamics. I approached Desmond to check on the story. He confirmed that it was true, and said the deal would be finalized by the end of the year. I told him I was now becoming extremely concerned and did not think I could, should, or would remain at our company. He again tried very hard to reassure me that everything would be OK.

Earlier in 1998 I had been in Seattle for a Boeing meeting and had run into one of the UK folks from GEC Marconi, Paul Childs. He was a marketing type for them, and despite the fact that we were fierce competitors, we had met many times at shows and other events and got along quite well. We’d even had a few beers together…carefully, like watching porcupines mate. He asked me if I would be attending the upcoming Farnborough Airshow, the very large and influential UK global aviation airshow that I had attended a few times in the past with Flight Dynamics. In fact, I had already been asked to go by Flight Dynamics.

“Why do you want to know?” I had asked Paul.

“Because,” he responded, “I want to introduce you to my boss, Dean McCumiskey.”

“OK…” I decided not to pursue it any further.

I happened to mention this to my colleagues during a subsequent watering hole event and they immediately became very alarmed that I might be planning a departure.

At the airshow in the UK, Paul came to our booth and told me he had set up a meeting with Dean at the GEC chalet. He proposed a day and time. Sure, I figured, what did I have to lose?

It was more than a little interesting to meet Dean and Paul at their chalet. Since GEC was a UK company, they had a huge and amazing setup on their home turf, and it was equipped with many of their products, including military HUDs and their VGS. They were both very complimentary about my successes at Flight Dynamics, and Dean asked me if I would be interested in a job with their company. I would be their Director of North American Sales for the new VGS. They explained that there were many corporate benefits to the title "Director." I told them I would think about it. I was certainly intrigued, and I would be able to continue my quest to change the way we fly airplanes.

When the Collins full acquisition was announced, I called Dean to ask if he was still interested in me, and he was. After some salary negotiations, I told him I would accept his offer of a job with GEC. We set a start date of January, 1999. He told me I would need to spend a couple of weeks at their facility in Rochester, UK, about 45 minutes southeast of London. No problem—I liked England.

I gave my resignation letter to John. He was shocked and extremely upset and asked what they could do to keep me. The A-Team also tried hard to convince me to stay, but I was determined. At a lunch at Wu's Chinese restaurant—one of our favorites—John and Al Caliendo, our CFO, told me they could give me a "retention bonus." I had never heard of that before, and although it was very tempting, I declined. My problem was not with Flight Dynamics.

I made plans to fly to the UK, rent a car, and spend two weeks with my new employer at their huge facility about an hour's drive from Heathrow airport. They had made a reservation for me at the Stakis Hotel, a Scottish Inn in Maidstone, a small village not far from their plant. There was a traditional English pub right next door called the Chiltern Hundreds, where I met up with Paul and Dean on occasion for some pub grub and beer. But I could not get used to the haggis offered at the Stakis for breakfast every morning...yech!

I was impressed by the GEC facility. They certainly had the credentials when it came to HUD. I made it very clear to everyone at GEC that I was not going to discuss anything about the HGS or my old alma mater's technology, and there was absolutely no point in asking me any questions regarding the subject.

An amusing thing happened on my first weekend in England. I had met Dean's secretary, Jan, on my first day, and she was very helpful. She even found me a vacant office where I could hang my hat for my initial visit. At the end of my first week of VGS education, and meetings with many key engineering and management folks, I realized I now had a whole weekend off. I asked Jan where I should go sightseeing. She took out a pad and pen and began to write, muttering to herself. Finally she looked up at me and smiled.

"Let's see," she said, "you can go to Leeds Castle, not too far away from here, or maybe go farther to Canterbury, a cute little Medieval town. Lots to see and do there. Or, there's Dover, quite a bit farther on, with the famous White Cliffs. You could also visit Folkestone, another neat little English village. Or, if you are really ambitious, you could drive all the way to Brighton, down on the coast," she finished, quite satisfied with herself.

"Wow, that's great!" I thanked her profusely, grabbed her list, and headed out on Friday after work in my rental car.

On Monday morning I returned to the office and got my coffee. Jan raced over to me and grabbed my arm, almost spilling my hot beverage all over me.

"Which of those places did you decide to visit?" she asked enthusiastically.

"All of them!" I replied. She looked aghast and almost dropped her own tea.

"What? Oh Phil, don't be daft, no one would do that."

Daft? I think that meant "crazy." Jan was absolutely incredulous that I would drive that far over just two days. I assured her that I had, and even showed her some of my hotel and restaurant receipts to prove it. Apparently that drive became the topic of conversation around the whole floor at the GEC office for the rest of the week. "That daft American."

In a later discussion with Dean in his office, I related his secretary's reaction to my travels.

"Phil," he explained, smiling, "in the US, a hundred years is a long time, but in the UK, a hundred miles is a long way."

We both laughed out loud, and I finally got it.

I soon realized that the GEC VGS was not that dissimilar from the Flight Dynamics HGS I had so much experience with. In a discussion with Dean a bit later, I mentioned that I thought my buddy Dick Hansen might be interested, and that GEC's VGS effort would be significantly enhanced with someone with Dick's talent and background working on their new project.

McCumiskey gave me the go-ahead to contact Dick. After I'd explained to Dick all about the VGS, the folks I'd met, and their facility, he decided that, given what was happening to Flight Dynamics, he too might enjoy a change. I connected him to Dean, and after some good discussions about roles and responsibilities, Dick agreed to become Director of VGS Training and Development – North America.

Soon after Dick started with GEC, we received a congratulatory telephone call from Brian Will and Rick Owen at American Airlines. Apparently, they had strongly recommended us to the management of their VGS supplier after watching us do our HGS thing during the American Airlines campaign. They had also witnessed our interactions with some of the Boeing folks.

Almost immediately, we uncovered two new opportunities for the GEC VGS. The first was Sun Country Airlines, based in Eagan, Minnesota, near Minneapolis. The airline had been started in the mid-80s by some former Braniff Airlines employees. In 1999, rumor had it that the airline was planning to order a small fleet of new B737NGs to replace a fleet of older

B727s.

The Boeing CHP project had made it easy for airlines with new B737 aircraft coming from the factory to select either the HGS or the VGS. One of my Boeing customer engineering acquaintances, who had told me about Sun Country's interest in new aircraft, also provided me with a contact name: TJ Horsager.

TJ was the airline's Flight Ops Specialist. I called him one day in mid-1999. He told me they had been made aware of the HUD option on their new aircraft, and he'd like to get a VGS presentation as soon as possible in Eagan, for him and some of the other key folks there. He was impressed that I had experience with both the HGS and the VGS.

The Sun Country office was located not far from where I'd had an "office" earlier at Northwest. Dick and I met a smiling and friendly TJ—another great aviation industry person—and he provided a little background on the airline. It was owned by the Mark Travel Group and controlled by the LaMacchia family. Bill LaMacchia Jr. was CEO of the airline. Their main focus was flying Minnesotans to warmer destinations during the winter months, such as Las Vegas, Florida, and Mexico, and doing charter work for the rest of the year.

TJ led us down a hall and into a conference room where we met more of the Sun Country team, including Dennis Vanatta, Director of Flight Ops, Brian Roseen, Director of Training and Standards, and Ron Payne and Pete Piazza, both Technical Pilots. After introductions and an overview of GEC, we presented VGS-101. As always, Dick did a great job with his pilot and HUD/HGS background, and we were able to answer virtually all the team's questions to their satisfaction. Dick and I stuck to our vow never to make any disparaging remarks about our old alma mater or the HGS in order to gain traction for the VGS in any airline HUD competitions.

We invited the Sun Country team to the M-Cab simulator to see and fly the GEC system, which was now also available for demonstration sessions at Boeing, and they accepted. Since many of the team had to attend some Boeing meetings in Seattle for final B737 negotiations and configuration, we timed the VGS simulator session for a slow period while they would all be there.

Meanwhile, during one of our subsequent Minneapolis visits, TJ introduced us to Tom Schmidt, their CFO, and Bill Hayes, their VP of Flight Ops, both of whom we knew would be key for a downstream decision in our favor. Bill especially seemed to fully support the idea, and became a high-level champion for us inside Sun Country.

A few weeks later, we met up with TJ, Dennis, Brian, and Ron in Seattle. After a great and quite lengthy M-Cab session—it was a big team—we invited the crew to Salty's restaurant in Seattle. We remembered what a hoot it had been at the chain's Portland location on the Columbia River with the

Delta team many moons earlier. Over a great seafood dinner and drinks, the group told us how much they had enjoyed the simulator session, and said that our VGS made "a good pilot great." We loved that line, and vowed to find a way to use it somewhere. It reminded me of Delta's argument about the HUD being an experience compensator.

While Dick and I were in the Seattle area, we decided to visit the GEC Marconi office in Redmond, set up there to support Boeing activities. At the office, we were introduced to Ron Barry, Glen Hislop, and Dave Kingstone—the same Ron and Glen we had seen during the American Airlines campaign in Dallas. We had fun with this bunch, with lots of chiding about the Scots (Glen was Scottish) and the Brits (Dave was British). They were very pleased with our addition to their team, and we had some good laughs about the final days on the American campaign trail.

A couple of weeks after Sun Country's visit to Seattle, TJ called me to announce that his airline had presented HUD proposals to their owners, who had approved the deal and left it up to Flight Ops to decide which system to go with. Flight Ops had selected our VGS, and we only needed to hammer out the terms and conditions. It was a small victory, but was key to proving to the UK management that our presence in the US could be a big asset to GEC's commercial HUD effort.

Dean had seen the HGS fat li'l aircraft stickers and asked me one day if I could come up with one for the VGS. Of course I could.

About this time, there was a lot of regulatory investigation into a 1998 Swissair MD11 accident off the coast of Nova Scotia, Canada. An electrical fire behind the instrument panel had caused heavy smoke in the cockpit. This was determined to be a big contributor to the fatal crash of the aircraft. I contacted Dean to tell him that I thought we should instigate a study to explore whether the VGS would have helped in such a scenario. My rationale was that with the brighter HUD symbology and the proximity of the combiner to the pilot's face, the VGS might have provided visible landing guidance after the head-down displays were no longer visible through the smoke. After all, the system was certified for landing in the murk. Would it work if the murk was inside the aircraft? The interior smoke issue had never come up while Dick and I were at Flight Dynamics.

Dean agreed and gave me the authority to look for a research facility that could conduct the study. After extensive effort and negotiations, I found Dr. Jim Blanchard, Professor of Human Factors at Embry Riddle (ER) Aeronautical University in Daytona Beach, Florida.

The ER lab was equipped with a smoke chamber that seemed ideal for our investigation. GEC supplied a VGS for the test, and the ER lab supplied standard PFD displays (Allied-Signal ED-102D). Two miniature video cameras were mounted in the eye holes of a ladies' styrofoam wig head. The head was installed in the typical pilot's head location. The cameras were independently connected to monitors and a VCR, which recorded data for subsequent analysis. Airline-issue smoke goggles (Scott PN 322-70) were supplied by an airline partner and fixed over the styrofoam head to simulate actual viewing conditions.

The styrofoam head, PFDs, and VGS were all mounted at positions and distances that emulated a typical B737 cockpit, even though the aircraft in the accident had been an MD-11. I was looking for more safety benefits for our system on the B737.

Smoke, of a density approximating that of electrical smoke, was introduced slowly into the chamber at various entry points. The cockpit instruments were set to nominal, and then to maximum brightness.

The test concluded that "the VGS commercial head-up display positively enhances flight critical cockpit instrument viewability during dense smoke conditions and provides flight critical information long after head down instrumentation becomes completely occluded by the smoke." GEC was to get a lot of good press over the results, and we had once again impressed our British overlords.

Dick and I visited Rochester, UK, on a fairly regular basis. After long days at the GEC plant discussing new VGS features and target airline strategies, we would end up at Chiltern Hundreds with Dean and Paul for a few frothy ones and lots of laughs.

On one occasion, Dick and I were in the UK with our wives, and found out that our good friend from Southwest, Paul Sterbenz, and his wife Pat were also there. We planned a get-together, and some folks at GEC recommended Ringlestone Inn, a pub built in 1533. Dick had offered to drive (on the left side). Our hotel gave us a map, but the inn was a bear to find. It finally required a lengthy drive through a field of grass and crops too tall to see over. The vegetation was hitting our car's mirrors on both sides and we worried about what would happen if we met another car coming the opposite direction. We finally made it to the charming historic pub for drinks and a delicious meal with our buddy from Southwest. There were lots of fun stories, jokes, and memories, and it was a great get-together.

Around this same time, we learned from another of our Boeing customer engineering buddies that two folks from South African Airways (SAA)—Brett Gebers, Chief Pilot, and Des Wise, Fleet Manager—were visiting Seattle to explore the available catalog options for their new order of B737NGs. Dick and I asked for their contact info and called to set up dinner with them. We got along well over a long evening of dinner and drinks, with us chatting up the many safety and operational benefits, both of HUD in general and the VGS in particular. We were now able to emphasize the benefits of the VGS "tray concept" that had worried me so much at Flight

Dynamics. After a great M-Cab session the next day, our new friends began in earnest to consider the GEC solution for their new fleet.

Shortly after he returned to Johannesburg, Brett contacted us and requested a technical briefing down in South Africa for the rest of their team, including some senior decision-making management. I was still up to my eyeballs with the final details of the Sun Country contract, so Dick volunteered to accompany our UK technical team to visit SAA.

I had previously scheduled a December vacation in Hawaii with my wife, but ended up spending a lot of that precious personal time on conference calls with Dean at GEC, and with Tom Schmidt at Sun Country, trying to sort out the final contract terms and conditions. Finally Tom told me we had a deal!

I called Dean to give him the news. He told me to take my wife to the nicest restaurant we could find in Hawaii, to help make up for the telephone interruptions on our vacation. We found a terrific seafood restaurant and toasted our new employer over a wonderful dinner.

I received a call from Dean shortly after returning from Hawaii, to tell me that GEC Marconi was about to merge with British Aerospace, another large UK aviation company.

There was quite a bit of "hush" about what the name of the merged company would be, but a short while later, Dean called again.

"Phil, guess what? You and Dick now work for BAE Systems." He laughed as he explained that this was the new name of the merged company.

"But don't worry," he continued, "you lads will not be affected except for some new business cards, letterheads, and envelopes."

That was fine with me. I briefed Dick on his return from South Africa.

A short time later, Dick was contacted by the SAA Boeing customer engineer to tell us that the airline had selected the GEC VGS and we had our second success. We had done it again! This time Dean authorized us to take our wives to a Ruth's Chris steakhouse in Portland for a celebratory dinner.

We were starting to really enjoy this.

BAE Systems' Electronic Systems facility in Rochester, UK

Wig head with goggles, installed in the ER lab smoke chamber. The test concluded that the VGS would be viewable long after the PFD was obscured by interior smoke, and could provide guidance for a safe landing.

GEC VGS fat li'l aircraft sticker

The Chiltern Hundreds old English pub in Maidstone, UK, was right next to our hotel, the Stakis

BAE Systems' Paul Childs, Chris, and me at Chiltern Hundreds having a pint in 2016

Dick Hansen, Pat Sterbenz, Chris, Jean Hansen, Paul Sterbenz from Southwest, and me at the Ringlestone Inn

Phil's Marconi and BAE business cards

(This and facing page) The BAE Systems VGS brochure, including the "tray concept"

Visual Guidance System (VGS)

The world's most advanced Visual Guidance System comes from BAE SYSTEMS, the leader in Head Up Display (HUD) technology. Having led the advancement of the HUD for 35 years, with more than 11,000 systems delivered, BAE SYSTEMS can now offer a system that introduces a new level of operational performance and safety features for executive, business and transport aircraft.

The VGS displays essential flight information to the pilot in his forward field of view, providing:

- Improved situation awareness
- Precise flight path guidance
- Reduced tail-strikes and hard landings
- Warning Alerts (windshear, TCAS and tail-strike)
- Lower take-off minima
- Improved energy management
- Take-off and roll-out guidance
- Unusual attitude recovery

The Future – Enhanced Visual Guidance System (EVGS)

Imagine being able to:

- Perform Cat III approaches at Cat I facilities
- 'See' the runway 2 miles out in zero-zero visibility conditions.

The future is closer than you think.

BAE SYSTEMS has successfully conducted proof of concept trials interfacing VGS with both millimetric wave radar and infrared sensor technologies which will further enhance operations.

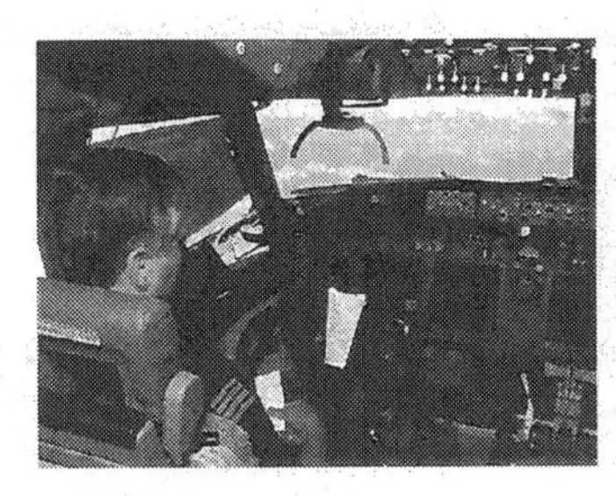

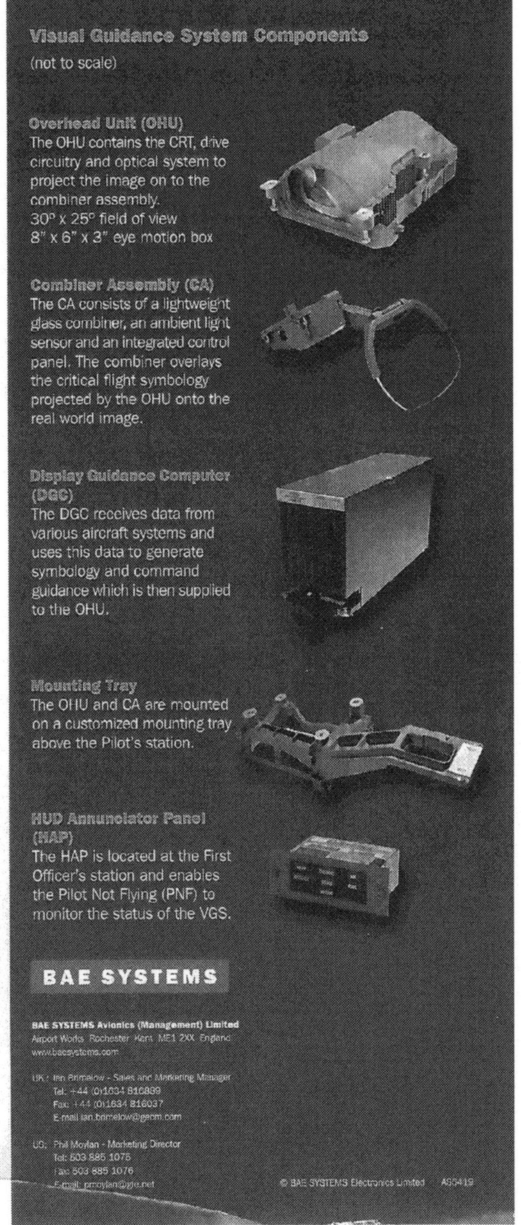

Chapter 23

Australia

Some very interesting things were starting to happen in Seattle. Boeing had decided, around the turn of the century, that they wanted to explore some new technologies for a planned upgrade to the 737 tentatively called the MAX. They had arranged a lease-back of one of Alaska Airlines' new B737-800 aircraft and were developing it as a "technology demonstrator." They would install new and developing technologies, and fly airline customers, regulatory folks, and even some media people, to gauge levels of interest and get feedback.

We found out that the Boeing Technology Demonstrator (BTD) program manager was Tim Tuttle. Dick and I met a very friendly Tim for lunch one day in Seattle, and asked him about getting our VGS on the aircraft. He said the Alaska aircraft already had the Flight Dynamics HGS installed. I should have known.

Believing that a HUD/EVS combo would achieve certification from the FAA for lower landing minima, Boeing added EVS to the aircraft to explore possible customer reaction. There were two suppliers: a small Oregon company called Max-Viz, and a Canadian company, CMC Electronics. Both of their systems were installed on the BTD. I clearly remembered the reaction to EVS that I had received in Asia after including the emerging technology in our HGS briefings there, and I suspected that most of Boeing's customers would also be very interested.

I'd been keeping an eye on my old alma mater, now called Rockwell Collins Flight Dynamics, and saw that they had developed their HGS for the newer Embraer ERJ-170/190 family of aircraft. Also, a very good customer, JetBlue, after ordering a hundred aircraft, insisted on dual HGS for their new planes. (Our previous Morris Air friend and HGS advocate, David Neeleman, was now running JetBlue in New York.) This dual HGS order was a first, and made me proud of my former successes with both Embraer and Morris Air.

I was still visiting Boeing in Seattle on a regular basis, and spending time with a lot of the customer engineers I had met over the years. Many of these folks I now called friends. In a discussion over lunch one day, one of them got my attention. Apparently, Qantas, in Sydney, Australia, was looking into a good-sized order of about forty B737NG aircraft. I immediately passed the info to BAE Systems management in the UK. Dick and I were told to head "down under" to convince Qantas to make the right choice. Neither one of us had ever been there, so this would be fun.

One of the benefits of the "Director" title at BAE was that Dick and I were our own travel authority—we did not need to ask permission to travel, anywhere. And for flights this long, we could, as directors, fly business class. Qantas was offering an amazing "free companion fare" for anyone flying business class that month, so we were able to bring our wives along. What a deal!

Since we were traveling such a long way, we had convinced McCumiskey at BAE that we ought to see other airlines while we were there. Ansett Airlines was located in Melbourne, and then there was Virgin Blue, a relatively new startup carrier headquartered in Brisbane. This gave us a few more potential targets as well as the chance to see more of their beautiful country.

So we headed down under to present VGS-101 to the good folks at our selected targets in Aussieland. On our visit to Qantas in Sydney, we immediately made friends with many of their personnel. There was Bruce Simpson, a B747 Technical Pilot, Bruce van Eyle, another Technical Pilot, David Oliver, GM Flight Technical, and Malcolm Campkin-Smith, their B737 Technical Pilot, who would be heavily involved in their new aircraft configuration effort at Boeing. All great folks. We made a lot of hay with the UK-Aussie relationship, and everyone was impressed that we had good answers for virtually all of their questions, even highly technical ones.

It was the first time Dick, Jean, Chris, or I had ever been down under, so we thoroughly enjoyed the sights and scenery of Sydney. We stayed at a downtown Marriott within walking distance of Darling Harbor, a great waterfront area with lots to do, see, drink, and eat.

The next meeting, with Ansett in Melbourne, did not go quite as well. The airline was obviously in serious financial trouble, and we believed they would either dissolve, or be gobbled up by another airline, in the very near future.

After our visit to Brisbane and a VGS-101 session there, Virgin Blue was interested, but also very preoccupied with setting up their new airline. It would likely be a while before they would become a viable customer, but they wanted us to keep in touch.

Our visit to Qantas had done the trick, though. Very soon afterward, the airline began asking Boeing how much for VGS provisions, and whether the VGS could be added to their order in time for first delivery of their new B737 aircraft.

Some of the Qantas folks came to Seattle for discussions with Boeing. We had a meeting, M-Cab session, and dinner with them, and they asked us to return to Sydney to present the VGS to some other key Qantas folks, including their Executive GM of Operations, and their Procurement Manager, both of whom would be critical for a go on the project there.

In the meantime, BAE Systems had a young and savvy technical engineer named Jason who had developed a unique VGS simulator that could be folded up into quite a small package for shipment to possible customers for system demos. Dick and I had played around with it during one of our regular visits to Rochester to familiarize ourselves with the setup and operation.

Once again, we found ourselves on a plane heading for Sydney, this time accompanied by a well-known HUD engineer from BAE, Paul Wisely, and our portable simulator. Paul had presented many white papers to the industry on military HUDs, and was now key to technical aspects of the commercial VGS. After arriving in Sydney, Dick and I met up with Paul for some strategizing over breakfast, and then headed to Qantas.

We set up our VGS simulator in a conference room at Qantas and were able to show a number of our Aussie buddies who had not made it to Seattle how the system looked and worked. Everyone was blown away. Gary Arndt, from Procurement, showed up and, although not a pilot, flew the VGS successfully. But we needed Mr. David Forsyth, the Executive GM of Operations, to come and see it. He was a key decision-maker, and the main purpose for our second Sydney visit. Apparently, he was in meetings until near the end of the day.

Our buddies at the airline continually visited his office, down the hall from our conference room where we were set up, to check with his secretary. But David's meetings just kept going on and on. It was getting late. Some of the Qantas folks were getting tired of waiting, and suggested we bag it and try again in the morning. But I was insistent that we hang on, since I felt that we might not get a second chance with such a busy executive. Besides, we would probably not be interrupted at this late hour, providing a good opportunity to brief him thoroughly without any typical office distractions. The others eventually agreed.

Finally, around 7 pm, long after secretaries and others had left for the day, our strategic guest got free. He was extremely apologetic for the delays. We got down to the business of showing him the VGS and allowing him to "fly" it. We were rewarded—he absolutely loved it, and was clearly enthusiastic about getting our system on his new Boeing aircraft. He flew it for quite some time before he finally had to leave for home.

Before leaving, he looked at us and asked, "Does Airbus know about your VGS?"

"I believe that BAE has shown the system to a number of Airbus engineering and flight ops folks in Toulouse, and I think they were impressed," I responded.

He then turned to his team, who were totally focused on his reaction and comments. "We also have ten or twelve new Airbus A330s coming, and we plan to order the new A380 double-decker aircraft. This VGS would be great for those aircraft too. We should be in discussions with Airbus about our interest in this technology for any new aircraft we buy from them, and we should do that very soon."

We were ecstatic, our Qantas team was grinning ear to ear, and I was so glad we had waited. What a reward!

After an incredible celebration dinner of wine and lobster on Darling Harbor with Wisely, we "folded our tent" and headed home, very happy campers. I could not wait to share our results with BAE's executive management—especially the Airbus discussion, and the Qantas desire to investigate the VGS for these aircraft also. Wisely told us he would brief BAE engineering management when he got back.

On a conference call the day after we arrived back in Portland, Dick and I discussed our down-under visit with BAE. While they were happy with our success there for the B737s, there was not much enthusiasm for undertaking a new VGS program for the A330 or the A380. BAE believed these Qantas fleets would be too small to justify the effort. They seemed unsupportive of system development and certification for new aircraft types. We pressed hard, trying to convince them that with a customer like Qantas, Airbus might be persuaded to consider the VGS as a factory option, à la Boeing B737, which would make it easier for us to go out and get other customers.

Three things happened almost simultaneously that would drastically alter our future at BAE.

First, American Airlines had enthusiastically received many VGS-equipped B737NG aircraft from Boeing, and were exploring the same system for about a hundred each of the B757 and B767 aircraft. BAE again showed resistance to VGS development and certification for new aircraft types. This really surprised us, since we had always assumed that their "tray" approach would make it easy to adapt the VGS to other aircraft types. Under their American Airlines contract, BAE had agreed that should the airline decide to move forward with other aircraft types, they would support. Now they began to waffle.

Second, our BAE contracts fellow was planning to fly to Sydney to close the B737 VGS deal and to tell Qantas management that BAE had no appetite for doing the new Airbus aircraft. We knew Qantas would not be pleased and we were concerned about the reaction, especially from Mr. Forsyth.

Third, and almost at the same time, 9/11 happened. BAE's management became convinced that the US airline industry would take eight or ten years to recuperate from this horrific event, and would not be looking at VGS-type technology for a long time. We desperately tried to convince them they were wrong, but no luck.

BAE also became concerned about using precious company technical resources on commercial HUD while the UK Ministry of Defense and the US government were now ramping up serious military spending as a result of 9/11. They decided they should refocus on their primary business and mission—military.

We were shocked!

One evening I was in Portland, enjoying a glass of wine with my wife after dinner, when the phone in my home office rang. It was our good buddy Bruce Simpson from Qantas.

"Oy, Phil," he started, in his now familiar down under accent, "I have an awkward question to ask you, mate. I was wondering if the Rockwell Collins HGS was as good as the BAE Systems VGS." Bruce knew Dick and I had spent many years with our old alma mater.

And I knew exactly where this was going.

"Absolutely Bruce," I assured him. "I would not hesitate for one minute to recommend their HGS, and the company." I knew that Qantas was very unhappy with BAE's decision not to consider the VGS for their new Airbus aircraft, and they were also hearing from Boeing that BAE was "pulling back" on the commercial air transport HUD business. Qantas was getting worried.

"Do you have someone I could call there at Rockwell?" he asked.

I provided Tom Kilbane's name and contact info, assuring him that Tom would certainly look after him. And I apologized profusely for the BAE reluctance to support their new Airbus fleets.

I mentioned this to Dick the next day and we were both very down. We knew in our hearts that this was likely the end of our association with the BAE Systems VGS. And it was—we received a telephone call a few days after 9/11 to notify us that our VGS sales and marketing services in the US were no longer needed.

But, as always, there was light on the horizon.

Boeing 737NG Technology Demonstrator (BTD)

Max-Viz and CMC Electronics EVS installed on BTD

A Qantas B737NG on delivery to Sydney

Dick and Jean Hansen, me, and Chris at Jordan's on Darling Harbor in Sydney, Australia

(This and facing page) BTD brochure

Boeing Technology Demonstrator

*Demonstration Flight Profile**

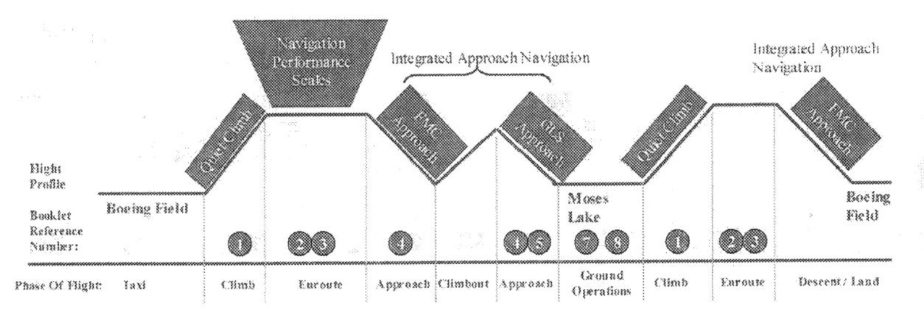

- The Enhanced Vision System (EVS) will be featured on all approaches to landing and during ground operations
- The Head-Up Display (HUD) will be featured on all approaches to landing
- The Synthetic Vision System (SVS) will be featured during all phases of flight
- The Vertical Situation Display (VSD) can be viewed from the flight deck during all phases of flight

*Because of Air Traffic Control considerations, this flight profile may be modified before or during the flight

Aircraft Layout

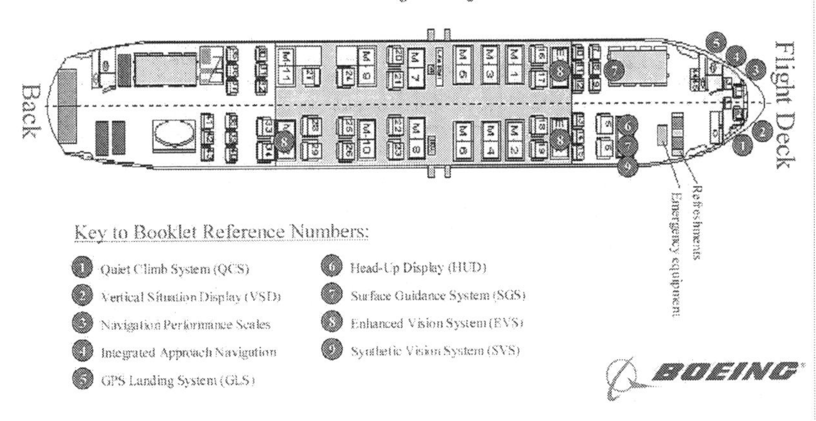

BOEING

Chapter 24

Icing on the Cake

Since leaving Flight Dynamics at the end of 1998, I had kept abreast of how my old alma mater was doing, and they were doing well...very well indeed.

Boeing had developed a new wide-body airplane, the B787, to replace the venerable B757/767. Rockwell Collins had managed to win the cockpit, and I was elated to find that dual HUDs were included as standard equipment on the aircraft. Boeing had always believed that both pilots should have all flight information. In a later discussion I had with some Boeing folks, I was told that the B787 HUDs were not there for poor visibility operations, but were mainly there for the many head-up safety advantages, and as a flight crew monitor during autoland operations. Also, with many emerging new avionics and cockpit features requiring display space, the HUD offered a good location for the aircraft Primary Flight Display. This freed up head-down real estate for new system display requirements.

After the very first flight of the B787, the test pilot was asked how the new cockpit displays had looked and performed. He replied that he had not really examined them very much because he'd conducted about ninety-five percent of the flight on the aircraft's new HUD. Funny!

Rockwell later managed to replace the VGS on American Airlines' new B737NG purchases. American had considered VGS for existing fleets of B757 and 767, but BAE was still refusing to adapt the VGS to additional aircraft types.

Later still, Rockwell also succeeded in beating out Honeywell-GEC on the newest members of the Gulfstream family of business jets.

Bob George, who had joined our marketing team before I left, was eventually given China, a big and challenging territory. But Bob did not disappoint. As I reviewed my monthly issues of aviation industry magazines, I was always thrilled to read that yet another Chinese airline had opted to take the HGS on new orders of B737 aircraft from Boeing. Later, I read that China was even developing a mandate that compelled all airlines flying in China to be HUD-equipped by the year 2020. This was truly an amazing development for international aviation.

At some point, one of my colleagues told me that virtually fifty percent of B737 aircraft leaving the Boeing factory now had HUD on board. Wow! We had, without any doubt, changed the way we fly airplanes.

For a long time I had been enamored with the Enhanced Vision System, or EVS, technology. I believed it was the consummate "icing on the cake" for commercial transport aircraft HUDs. The HUD-EVS combination had been in use for many decades on military fighter aircraft, and a lot of the same benefits were recognized and understood by the commercial aviation industry. Many runway incursions and CFIT (Controlled Flight Into Terrain) accidents had happened in darkness, and the infrared-based EVS would have provided a daytime-like image to the flight crew.

Gulfstream's pioneering EVS efforts in the late 1990s led to FAA certification for lower landing minima in October, 2001, and this opened the flood gates. The Canadian company, CMC Electronics, won the EVS business at Bombardier for their larger bizjets (the GEX and the G5000). Rockwell Collins worked closely with CMC to develop EVS for the Rockwell HGS on the family of Dassault bizjet aircraft and for BBJ.

Much later still, Rockwell developed their own EVS. I distinctly remembered the reaction to the EVS images on the China trip with Collins. I am convinced that, eventually, Chinese aircraft with HGS will also be equipped with EVS.

CMC and others began looking into integrating other forward-looking technologies, such as Millimeter Wave Radar/Sensor and LIDAR, to develop a Combined Vision System (CVS) that would provide even more capability in poor visibility conditions.

Meanwhile, commercial aircraft HUDs were beginning to pop up everywhere.

Elbit, an Israeli company that was originally an exclusive military HUD manufacturer, got into the commercial HUD business and eventually won the newest Dassault F5X and F8X business jet programs with a fully-integrated, single-supplier solution they called FalconEye. They came up with their own five-sensor EVS to support their HUD offering, claiming many additional benefits.

Thales Avionics, formerly Sextant-Avionique, became the sole HUD supplier to the Airbus family of aircraft, and Garmin, a well-known and relatively inexpensive cockpit display supplier for the smaller and mid-size bizjets and turboprops, also developed their own HUD. A small company, MyGoFlight, by modifying automobile HUD technology, came up with an inexpensive HUD for very small General Aviation aircraft.

Not to be outdone, AVIC in China teamed up with GE Aviation to develop a HUD for the new Chinese airliner, the C919, and their proposed future wide-body, the C929. Their HUD's optical portion was developed by LIEOE, and the computer by SAVIC, both Chinese companies. Saab Avionics, one of the newest entrants into the HUD world, developed a unique "synthetic vision system" (SVS) cockpit display that included their own dual

HUDs. (More about SVS in a minute.)

No question: things were really looking up!

In 2003, I formed my own aerospace marketing consulting business and my great career buddy, Dick Hansen, joined my new endeavor. My first real client was CMC Electronics in Canada. My background in HUD and EVS was a big help to them, and working with them kept me current in the world of commercial HUD.

A few years after starting my business, I was contacted by my good BAE friend, Paul Childs.

"Hey, Phil, would you be available to come to BAE Systems in Rochester for a meeting with us?"

"Is this about commercial HUD?" I asked him. I'd heard they had decided to get back into civil HUDs.

When he said yes, I replied, "When do you want me there?"

After flying to London and driving to the old familiar BAE facility, Paul met me in the lobby and escorted me up to a conference room where a number of folks, some of whom I knew from my early VGS days, were present. I had peppered Paul about the purpose and objective of the meeting but he'd told me I would find out soon enough.

There were a lot of familiar smiles and welcoming words around the table. Then they explained that they had developed a whole new HUD technology that eliminated the need for the traditional overhead projector unit. The size and weight of the projector had tended to eliminate mid-size and smaller bizjets from the world of HUD. I was all ears!

To eliminate the projector, they had recently developed an "optical waveguide" HUD they were calling QHUD. It used a revolutionary means to "inject" the symbology into the combiner rather than projecting it onto the glass screen. This resulted in a HUD that could fit into much smaller cockpits that had very little room above the pilot's head, such as the Cessna light, medium-size bizjets, and the Bombardier Learjet family. It even had possibilities for aircraft like the Beechcraft King Air, many of which were being delivered from the factory in Wichita as Special Mission Aircraft where a HUD was really desired. BAE Systems needed my help in North America. I was thrilled to once again be involved in new commercial HUD developments.

One of the most unusual results of the new BAE technology was that the combiner no longer had to be tilted upward to reflect the overhead unit's projected symbols, but could be vertical, or even tilted the other way, so that it was now parallel to the aircraft windshield. I told them I wanted to see it.

They took me to a different area on the same floor and showed me the QHUD prototype. I was invited to sit behind the combiner. The engineers explained that this first example was smaller than the production version would be, and yet the "eyebox," or viewable area, was huge—in fact it extended back from the combiner in a column the same size as the combiner itself. I could scoot back two or three feet from the combiner and still see the symbology!

After I asked a few more questions, I suggested we stay away from aircraft platforms that already had a HUD or an established HUD supplier. We should chase the "virgin territory" where there were lots of small to medium size business aircraft, and maybe even some larger turboprops, that had no HUD, and where conventional HUDs would have a hard time fitting into the cockpit. Everyone agreed.

Now under contract with BAE Systems through my own company, and back in the US, I set up QHUD presentations at Cessna, Beechcraft, and Learjet, all in Wichita. Three folks from the UK supported me at these meetings: my buddy Paul Childs, from Business Development, Andy Thain, the QHUD Program Director, and Fraser McGibbon, from Technical Engineering Support. The four of us got along famously, and over the next couple of years we spent many days and weeks together educating the business jet community on BAE's newest HUD technology and its benefits. We also attended airshows like the NBAA, and the EBACE in Europe, where we exhibited the new optical waveguide HUD technology and attracted a lot of industry attention. It brought back memories of my first day on the job with Flight Dynamics many years ago, in 1985.

But the technical group in Rochester had run into a snag. There were some optical "noise" artifacts showing up as "spackle" on the larger combiners needed for the mid-size bizjets, and the engineers were having some difficulty eliminating the anomaly. But the team was intent on overcoming the obstacles, and so they trucked on. I was asked to hold on while the issues were sorted out.

A funny story happened when a group of my BAE buddies came to Portland to visit some Oregon bizjet operators. One of the guys saw an ad for a small town nearby called "Wankers Corner." The whole UK team erupted in laughter, as the word wanker had a very different meaning in British English. They insisted I take them to the village, where they piled out of my car and had me take photos of them in front of the village's large welcome sign. Then they descended on the small convenience store, which had T-shirts, baseball caps, coffee and beer mugs, and other items emblazoned with the town's name. They bought quantities of stuff—one of the lads picked up eleven T-shirts!

While I was waiting for BAE to sort out their technical issues, Rockwell Collins was generating lots of industry press over their inclusion of Synthetic Vision System (SVS) technology on their HGS. SVS was a term used for computer generation of images of the scene in front of the aircraft, derived from satellite terrain databases. Knowing the exact GPS location of the aircraft and its altitude, the system was able to construct, from the database, an artificial forward scene like the pilot would expect to see through the windshield. The technology had been in use for some time on bizjet PFDs, replacing the traditional "blue and brown" displays.

Adding SVS to a commercial HGS was tricky, because the image could not obscure the actual view through the windshield. The Rockwell A-Team had done an amazing job as usual, and it began to look like SVS, as well as EVS, would become integral to the next generation of commercial HUDs.

About the same time, Rockwell also garnered a lot of attention for a new optical waveguide HGS they dubbed the HGS-3500. With its extremely small size and light weight, it was selected for the newest mid-size Embraer Legacy 450 and 500 bizjets. Bombardier jumped on it for their new Learjet—most of their other aircraft in production now offered the HGS, and they felt they could only tout the many safety benefits of HGS if all their aircraft had that option.

Rockwell went on to develop and supply their HGS for FedEx's B757, B767, and even B777F aircraft. FedEx had long been a fan of the HUD technology. And much later, Rockwell convinced Boeing to offer their latest EVS as well as dual HGS as options on the newest B737MAX aircraft.

HGS was now a flight deck option on virtually all Boeing aircraft in production.

The well-attended international Farnborough and Paris Airshows were quite interesting. For many years, Airbus had used these events to fly their newest aircraft—often very close to the allowable flight envelope. Their auto-systems flight envelope protection prevented any exceedances that could be deemed unsafe.

During that same timeframe, Boeing had opted not to fly their aircraft, since they were unsure whether it actually assisted new aircraft sales. But with their latest B737MAX rolling off the production lines in 2018, they decided it was time to fly at the airshow. The selected pilot said that the aircraft he was to use at the show *must* have the HGS, because it provided the kinds of envelope indications that would be very useful in any of his aircraft's extreme maneuvering demonstrations for the event.

There was absolutely no doubt in anyone's mind that Flight Dynamics, Inc., the small company in Oregon, had permanently changed aviation…for the better!

flightdailynews.com

HEADLINES

BAE gives pilots more HUD-room

BAE Systems' new Q-HUD, unveiled this morning, is expected to make the head-up display (HUD) a more practical display option for business jets by solving cockpit space and weight challenges. This next-generation HUD is scheduled to enter service in 2010.

"Q-HUD is very radical, very disruptive, very different from any conventional display," says Paul Childs, Q-HUD technology lead for BAE.

DEVELOPMENT

Ric Morrow, director of commercial avionics business development for the UK aerospace giant, adds: "Airline and military aircraft have used HUDs for the last couple of decades. Now we can expand the technology to bizjets. With Q-HUD, business aviation has access to a HUD that will fit in the cabin."

Unlike the traditional projection-style HUD positioned close above the pilot's head, taking up cockpit real estate, Q-HUD eliminates the need for the overhead box by generating images within the HUD glass using holographic waveguides. Instead of shining an image on to the surface of a combiner from behind the pilot's head, Q-HUD injects the image through the side of the glass.

By alleviating the need for projection, this unique design allows the unit to be mounted above the cockpit's closeout, so the combiner is the only visible component of the system. In addition, the pilot's head will not block the projected image – a problem that occurs with conventional HUDs.

While pilots will still see basically the same display, Morrow says Q-HUD is 50% smaller and lighter than traditional HUDs, providing increased head clearance and comfort in the cockpit. This space savings is the key that allows a HUD to be deployed into most sizes of business jets.

Additional advantages include economical price points for business aviation; weight reduction, compared to the heavy optics equipment of typical HUDs; and reduced power consumption and heat generation.

"We were able to improve the technology and reduce space, weight and cost, and we didn't have to compromise on performance or reliability," Childs adds.

Q-HUD can integrate with enhanced or synthetic vision interfaces, and existing flight deck computational resources, potentially saving the cost of a separate display.

Although BAE has not secured a launch customer yet, the company is in discussions with several airframe manufacturers and Tier 1 suppliers.

JUMBO

For now the company will focus on aircraft ranging from medium-sized business jets to jumbo passenger aircraft, but Morrow believes Q-HUD is a technology step that will take BAE toward the VLJ market as well.

To date, BAE has only manufactured HUDs for three civilian aircraft types – the Gulfstream GIV and GV and the Boeing 737-800 – but Morrow is still confident Q-HUD will gain wide acceptance.

"We have done some preliminary briefing with some very targeted audiences, and the technology has been very well received," he says.

At NBAA, BAE will be demonstrating Q-HUD technology alongside Q-Sight, a helmet-mounted display launched by BAE last year for the military rotor wing sector.

This Q-HUD article appeared in several aviation industry magazines and press reports

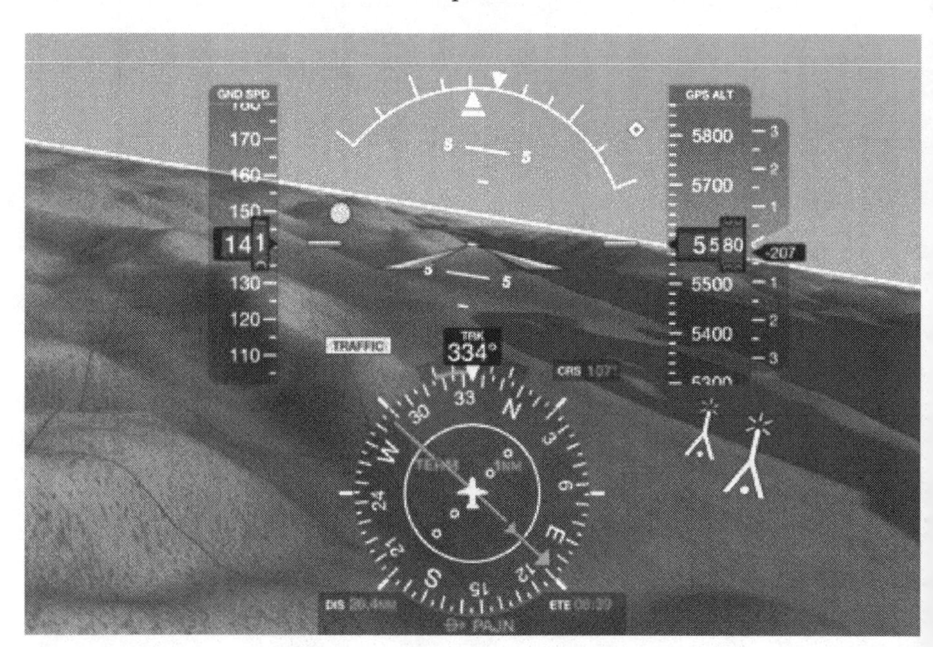

SVS on a Primary Flight Display

The SVS "ghosting" image overlaid on the HGS

B787 dual HUDs

Flying Collins' HUD with Synthetic Vision

Raising the bar for situational awareness in IFR conditions

By Fred George
fred_george@aviationweek.com

Next year, Bombardier's Global 5000 and 6000, fitted with Global Vision cockpits powered by Rockwell Collins Pro Line Fusion avionics, will be the first production business aircraft to enter service with head-up displays with synthetic vision system (SVS) background imagery.

Their HGS-6000 HUDs will use the same digital terrain elevation database (DTED) as the head-down primary flight displays, but the 3-D imagery will conform to the view of the outside world through the HUD on a clear day and the SVS picture will be monochromatic green to provide contrast with the rose-tinted combiner glass. Global Vision's SVS imagery also will be displayed on the head-down PFDs in full color and with a much wider angle field-of-view.

We had the opportunity to fly the Global Vision HUD with SVS in late September. It was installed on the right side of the cockpit of the Rockwell Collins Challenger 601-3A flying test bed, along with a full set of Global Vision flight instruments including a 15-in. PFD. Synthetic vision rapidly is emerging as a de facto standard for head-down displays in business aircraft, so a logical next step is to move it up to HUDs, according to Rockwell Collins engineers.

HUDs with SVS have been in use aboard military aircraft for more than a decade, especially tactical aircraft that fly high-speed, nap-of-the-earth sorties. We've witnessed the capabilities of just such a system while flying a low-level demo hop in a Dassault Rafale in central France.

Now, Rockwell Collins is pioneering the use of SVS for HUDs in civil aircraft. The goal isn't to give civil pilots the ability to use SVS to fly down canyons and skim over ridgelines in IFR conditions, but to enable them to fly down to lower minimums on published instrument approaches in accordance with FAR Part 91.175, much the same they can in some aircraft equipped with HUDs having approved enhanced vision systems (EVS).

We've flown several aircraft with HUDs having EVS. These systems use external sensors, typically short-wave infrared cameras, to peer through clouds and darkness to detect terrain contours, electric lights and runway surfaces. What they detect then is projected as background imagery on head-up or head-down displays to aid pilots in "seeing" the runway environment better than they can with unaided vision.

One of EVS's key assets is its ability to detect obstacles in the landing zone in real time. If, for instance, a vehicle, aircraft or animal moves onto the runway, an IR EVS may be able to detect the threat beyond the range of unaided vision. The HUD then

SVS provides a 1-km radius airport dome symbol to help pilots locate the landing facility well beyond visual range, even in clear weather conditions.

Rockwell Collins

SVS technology made its way onto the Rockwell Collins HGS

The Garmin HUD with SVS in a glass cockpit

The small optical waveguide Rockwell Collins HGS-3500 with SVS

Saab Avionics gets into the mix with SVS (l) and CVS

Epilogue

Looking Back

It is hard to fully describe, as I write this book in 2018, just how rewarding and enjoyable my decade-and-a-half career at Flight Dynamics was. The often difficult and seemingly insurmountable challenges that Jim Gooden had originally revealed to me, we had faced head-on and overcome. And the quality people I met, both at our little Oregon company, as well as at the customer organizations that I chased after, made me realize what a special industry this is to work in. Other than playing in a rock band in my late teens and early 20s, and a summer job in architecture, I never worked a single day outside the aerospace industry—and I have loved every minute of it.

I cannot express how grateful I am to have been such an integral part of the Flight Dynamics team and the company's eventual successes. We shared a lot—struggles, wins, losses, challenges, jokes, downers, fun, beers, but in the end, we did it.

We changed commercial aviation…for the better!

How I Got into Aviation

During my years with the Flight Dynamics A-Team, and with the many great friends and acquaintances I made at the airlines and OEMs, I was often asked, usually over a few beers, how I got into aviation, and how I ever ended up in sales and marketing.

Here is the story of what led me to my aerospace career and the eventual writing of this book.

How did I get into Aviation?

In 1967 I graduated from St Patrick's High School in Ottawa, Canada, at the age of 19, and enrolled in an architecture course offered at nearby Algonquin college. I enjoyed my first school year designing houses, office buildings and, in one case, a lavish seaside vacation resort. My father, who was in the commercial fire alarm business, managed to get me a summer job with a local architectural firm through one of his contacts. I absolutely hated it, and realized that, in the real world of architecture, only ten percent of the time did you actually get to design the building. The other ninety percent was spent on plumbing, electrical, lighting, power outlets, toilets, and ensuring compliance to a myriad of often ridiculous government regulations. At the end of the summer, I told my Dad that I did not want to go back to architecture at Algonquin.

He told me that if I was not going to school, I would need to contribute to the household.

"What do you mean contribute?"

"We think a hundred dollars a month should cover it," my father replied.

"A hundred dollars a month!! Are you kidding? For that much money, I could get an apartment," I almost shouted at him.

"Riiiight." He grinned at me.

So I started looking for a job. My dad told me to get out the telephone book yellow pages and search for something interesting, an approach I thought was absolutely ridiculous. However, to get him off my back, I took a look. I soon figured out that it was not such a bad idea.

One of the very first companies listed was Aass Aero Engineering Ltd., an aviation engineering company. Hmmm, I thought, airplanes, that sounded really interesting. Their office was located at the Ottawa airport, about three miles from my folks' house in Alta Vista. I hopped into my POS car, a rusting old 1959 British Ford Prefect, and drove to the address listed in the phone book. Entering the office, I was greeted by their secretary, whose desk nameplate said Netta.

When I told her I was looking for a job, she shrugged, held up her index finger, and disappeared into an office. A few minutes later, an older gentleman emerged, came over to me, and introduced himself as Nat McCormick, the company's Chief Engineer. Nat explained that they were a very small aeronautical engineering outfit developing aircraft modifications. The company comprised about seven people, and all positions were currently filled. After noticing some cool aircraft drawings lying around, I told him I had good drafting experience, which I had picked up over my brief architectural career. Still, he insisted, they had no need.

I was discouraged.

That evening at home over dinner, my dad asked how my first effort had gone. I told him about my visit to the aviation engineering company at the airport.

"Do you think you'd be interested in working there?" he continued.

"Yes, I sure would. I saw some of their drawings of aircraft modifications, and I think it would be a lot of fun. But they have no need of additional help."

"Well, if you think it's something you'd like to do, you don't want to give up," he said. "Go back tomorrow, and this time, bring along some of your architectural drawings to show them." Great idea!

Back at the Aass Aero Engineering office the next morning, with some of my rolled-up drawings under my arm, I told Netta that I wanted to show Nat my work. He had not yet arrived, so she went into a different office and another gentleman came out. Netta introduced him as Haakon Aass (a Norwegian name, pronounced Oken Oss), the company founder and president. We shook hands, and I opened my roll of drawings. He seemed impressed and asked some questions, but again, reiterated that they had all the help they needed. He promised to keep my skinny resume on file.

More discouragement…

Back home at the dinner table that night, my dad was anxious to find out how things had gone. I mentioned meeting the company president and that, while he'd been impressed with my school drawings, they simply had no openings.

"So, are you going to give up?" he queried.

"Well, yeah," I replied, "what else can I do?"

"Would you really like to work there?" he asked.

"Of course, but what the heck can I do if they don't need any help?" I was getting quite frustrated with this ongoing dialog.

"Go back and tell them that you will work at their company for no salary until you are carrying your weight," he told me.

"What? Are you crazy? Work for no money?" I was incredulous.

"Yes," was all he said, and went back to eating his dinner. I could not believe that this was his idea of "valuable job advice."

But, I thought, what the heck have I got to lose? They can't do any more to me than they have already.

Poor Netta just burst out laughing when I walked back into their office the next morning. But I held up my hand and told her that I had a great idea, and an offer they simply could not refuse. Nat was in the office already. He invited me into Haakon's office and they both sat down smiling to listen. Feeling like a total idiot, I suggested that I was willing to work there for nothing until such time as they both felt I was earning my keep. They looked at each other, and Nat told me to wait in the lobby while he and Haakon discussed my offer. Well, maybe that was progress…

A few minutes later, they invited me back in. Haakon told me that, while they were very impressed with my persistence, and especially my most recent offer, they could not accept, since it was illegal in Canada to hire someone and not pay them. I tried to push, but got nowhere. I left their office with a clenched fist intended for my dad over dinner that night.

I drove home muttering angrily to myself, parked in the driveway, and walked in the front door. My mom was standing at the head of the stairs into the kitchen.

"They called," she said, "that company that you've been going to see at the airport. They said you can start on Monday, and something about paying you as well."

Holy crap! It had worked! I unclenched my fist. That night over dinner, my dad congratulated me and shared his philosophy on life.

"Whether you believe you can or you can't, you are right!" he said. I never forgot that.

My work at Aass Aero Engineering entailed developing detailed drawings of aircraft modifications that Nat was designing for some geophysical aircraft customers near the airport. We did all kinds of helicopter and government aircraft mods to install large aerial cameras, tail booms, and sensor systems. Haakon, as a Design Approval Representative (DAR) for the Canadian Department of Transport, could stamp and approve the mods. I learned a lot, and quickly. I was soon suggesting changes to the way we approached these aircraft modification projects, and both Haakon and Nat were impressed with my ideas, and my eagerness and progress in learning about the engineering aspects of our work.

One day, about a year after I started, Haakon invited me out to lunch. I sincerely hoped it was not to fire me. Over our meal, he complimented me on my efforts to improve our company's efficiency, and asked if I'd be interested in engineering courses at a local university, Carleton, about five miles away.

When I said yes, but I needed the job also, he suggested that they could provide me with flex time to take some courses, but I would need to make up the time over evenings and weekends. I jumped at the chance.

It took me over six long years—including classes every summer—to finish the four-year course in aeronautical engineering, even with college credit for one of the architectural courses I had taken, and for a project I'd worked on at Aass. But I did it, and graduated in 1978. I was now officially an aeronautical engineer.

I was the first student ever to complete the course part time (see the article from the Carleton University graduation newspaper at the end of this chapter). Immediately afterward, I applied for, and with a great endorsement from my DAR boss, received my own DAR appointment from the Canadian Ministry of Transport. I was DAR-87. This is equivalent to the US FAA DER appointment.

Whenever I considered post graduate work, I would open my third year Thermodynamics textbook to page 289. There was a hole thru about eighty pages made by a BIC ball-point pen, with blue ink and dried blood all over the page. It had happened while I was studying for my finals that year. I still have the scar on my hand!

How the hell did I get into sales and marketing? It's a funny story…

About two years after graduating, I was offered an engineering job at another local Canadian company, Leigh Instruments Ltd. It was a much bigger company, with four or five hundred employees. Their claim to fame was a family of aviation crash position indicators and recorders, the little "black boxes" that everyone looks for after an airplane accident. Leigh was a pioneer in the development of the technology.

I would be their project engineer for a different product, their Mechanical Strain Recorder (MSR). This device was used to monitor structural metal fatigue, a phenomenon that, if not controlled, can lead to catastrophic structural component failure, often resulting in fatal aircraft accidents. This unique product was a self-contained unit that did not require electrical power. It used the strain energy of the member to which it was attached to advance a small foil tape on which was scribed the actual measure of the strain amplitude and number of cycles. A large computer called the DTU (Data Transcriber Unit) could then be used to "read" the MSR's foil tape and assess the fatigue damage.

Fatigue cannot be measured directly. It's like an opaque wine bottle—the only way to know how much remains is to monitor what you take out. The MSR did that. The company had been successful with a win for a new USAF fighter at the time, the F16, built by General Dynamics in Fort Worth, Texas. The MSR monitored main wing fatigue usage during the aircraft's extreme aerial maneuvering requirements.

A few months after I started, I saw an announcement in a popular industry aviation magazine for a fatigue conference sponsored by ICAF, the International Conference on Aeronautical Fatigue. It was to be held six weeks later in Brussels, Belgium. I wrote to the sponsors explaining, at a high level, our MSR product and benefits and asking whether, if we supplied a bunch of brochures, they could place them on a table at their conference.

Two weeks later, I received their response. In the letter, they talked about the conference agenda, their sponsors, and the pedigree of the attendees, and said, "Your white paper has been accepted for presentation. We will contact you soon with date and schedule and time allotment for your paper."

I went to my new boss, Mike Kwong, the VP of Engineering at Leigh, and showed him the letter.

"Why, this is *great* news, Phil," he almost shouted, as he reached for the telephone to call senior management.

"What the hell is a white paper?" I asked him, utterly confused.

"It means they will allow you to present the MSR at their conference," he replied.

He was ecstatic, and assured me that Leigh's management would pull out all the stops to support my effort at the conference.

"Do you have a passport?" he continued.

"No," I replied.

"Well, go get one right away." He laughed heartily as he returned to the phone to inform upper management about my white paper.

What had I done?

The next few weeks were totally occupied with preparations for my white paper. I needed a carousel of slides (that was how we did presentations back then, before PowerPoint) and had to go over each one of them, often with our engineers, until I got the story and explanation exactly right. Sometimes they would ask questions, simulating what I could expect at the conference, and correct my answers until everyone was satisfied. They assigned one of the Marketing Managers, a guy named Jim, who looked after the crash recorders, to accompany me to handle any questions on cost or product delivery schedule. And, of course, to pay our expenses.

The big day for our departure finally came. I was born in Dublin, Ireland, and had come to Canada with my folks when I was just twelve years old. I had never been back to Europe. This was going to be fun.

Jim and I flew to Montreal, where we boarded an Air Canada international flight. We arrived in Brussels, checked into our hotel, and went to the conference to register. They respectfully reminded me that I had forty-five minutes for my presentation the next day, and only ten to fifteen minutes to address any audience questions.

After a short snooze at the hotel, Jim and I walked to the Grande Place, the central square in Brussels, for a great dinner and some drinks. It was really fun! I was even able to use a bit of the French I had picked up playing in a rock band in the bilingual Ottawa area. (My band had actually played for a Leigh Instruments Christmas party.)

The next day, everything went well. I presented the MSR to a receptive audience of a hundred and fifty to two hundred industry experts on aircraft fatigue from all over Europe. I was asked a number of questions afterward, and luckily was able to answer them all. I was thankful for all the prep work we had done back at the plant. Jim and I distributed a lot of MSR brochures with my business card attached, and I felt sure we would get some serious interest in the product.

That evening, as we were preparing to go out for a celebration dinner, one of the VPs of Leigh Instruments called Jim in his hotel room. As we walked to dinner at the square again, Jim told me what had happened.

The VP had asked how my MSR pitch had gone, and Jim confirmed that it had gone well. The VP then told Jim that we were not flying home the next day as planned, but were to come to the UK to give a presentation during "Canadian Aerospace Technology Days" at Canada House in downtown London. At the last minute, our company had been given the opportunity to attend and present. Representatives from a large portion of the UK aviation industry, including the government, airlines, aircraft and helicopter operators, OEMs, and all kinds of suppliers, were scheduled to attend. Leigh was given an hour and a half from 1:30-3:00 pm on the first day of the two-day event. The VP would give a short executive overview of the company and history, Jim would cover the crash recorders, and I would do the same MSR presentation that I had done in Brussels. It sounded simple, but Jim looked really nervous, and he didn't want to talk about it.

We arrived in London quite late and hit the hotel. The next morning, we took a taxi to Canada House. As we entered the main conference auditorium, Jim started to freak out.

"What's the matter, what's wrong?" I asked.

"I...I can't do this," he stammered. "There's about a thousand people out there!"

"So...haven't you done this many times before?"

"Well, yeah! But only to seven or ten people at a time. I've never gone up in front of that many people before. I can't do this." Before I could say anything, he marched off to find the company VP.

A short while later, I saw the two of them talking, and I could tell it was getting quite heated. There was arguing and finger pointing and head shaking. Finally Jim walked off and the VP called me over. He told me he had just fired Jim because he refused to do the crash recorder portion of the presentation. He had brought the appropriate slide carousel from Ottawa, and now, he handed it to me.

"Phil, I need you to present the crash recorder portion since Jim is right now heading to the airport for his flight home," he told me.

I was dumbfounded.

I knew very little about our company's main product line. Oh, I had seen some of the units lying around our plant in various stages of assembly or repair, and had asked a few questions, but I did not know nearly enough to carry on a technical presentation for thirty minutes. But the VP was insistent. He would do the corporate intro, with background history and company financials, but he wanted an engineer to do the actual product show and tell.

How did I get myself into these things?

I took the slide carousel into a nearby men's room—the only place I could set up the borrowed projector—and reviewed the pitch. As I went through the slides, I removed any that I didn't understand, or didn't know what to say about. When I was finished, I only had about ten of the original thirty or thirty-five slides left. I became extremely concerned.

But I came up with an idea.

Going to the kitchen at Canada House, I asked the chef if I could have an egg. He asked if I wanted it boiled, scrambled, or fried. I told him uncooked. He looked at me funny, but gave me a raw egg.

Finally, at 1:30 pm, after the audience had all filed back into the main conference room after lunch, the VP got up to the podium and provided our company introduction, overview, and financials—eight or ten slides. Then he introduced me and sat down. I had decided to do the MSR first, since I knew what I was doing with that. The thirty minutes seemed to go by in a flash. I appeared to get some good audience reaction, and was able to answer all of their questions.

But now came the crash recorder. God help me! First, I told a joke—two or three minutes burned maybe. Then I removed the raw egg from my pocket and held it up for all to see. The VP, now sitting back down in the front row of seats, looked on in surprise and wide-eyed concern.

I began by explaining that the egg's shell was designed by nature to protect the vulnerable payload…the chicken embryo. I held it between my inverted palms with intertwined fingers and gently tried to crush it, explaining that the shell was very strong in its protective role. After extolling the structural virtues of the eggshell for a few minutes, I suddenly turned and hurled it against the concrete wall at the side of the stage. During the conference lunch break, I had taped some newspapers to the wall and placed some more on the floor below. The liquid and broken shell pieces from the raw egg slowly dribbled down the paper.

You could hear a pin drop in the room. The VP was agape…

"It's not pleasant to think about," I said, "but that's what happens when mass and inertia exceed the allowable stresses of the shell, and that's what happens in a plane crash."

"Now, if someone new enters the back of the hall and sees the dribbling egg, they might assume that someone did not like my joke, or was trying to get my attention. But until they sit down and ask the person next to them what happened, they do not know for sure." I was on a roll.

"That's what our crash recorders do," I explained. "They tell the story after the event has occurred. They help determine the cause, and prevent other similar accidents from happening."

I now went on to the ten slides that I knew a little about (from product brochures the VP had given me), and managed to hack my way through to the thirty-minute deadline. A few quick questions and I was done. Whew!

As the crowd cheered, I gazed down at the VP, who was grinning from ear to ear and appeared to be very pleased with my "egg show."

Afterward, as I was gathering my notes and slide carousels, he approached me from behind.

"Hey, Moylan." I almost jumped out of my skin, fearing the worst.

"Guess what?" he asked.

"What?" I asked, nervously.

"You are now officially in marketing at Leigh." He laughed. I was stunned.

I guess I was lucky. In our rock band, I somehow ended up doing most of the talking between songs, so I had a lot of practice in front of sometimes very large audiences in the clubs and high schools and occasional giant stadiums where our band had played.

Later, at London's Heathrow airport for my return flight to Canada, I noticed on my boarding pass that I had been upgraded to first class! I was seated next to the VP. He asked me where the hell I had gotten the idea for the egg. I told him that it was desperation. I needed some filler, but also something that was hopefully relevant to what I was talking about. Laughing heartily, he complimented me and said that he would speak to my boss, Mike Kwong, and let him know that I was being moved upstairs to Jim's office.

And that's how I joined the sales and marketing department of Leigh Instruments.

I left Leigh a few years later, after receiving an offer to manage the Simmonds Precision Canadian office based in Toronto, and the rest of my story is told in this book.

In June of 1983, sometime after I left Leigh, there was a terrible Air Canada DC-9 accident. Some Leigh employees were coming home from an F16 MSR meeting with General Dynamics. About halfway through the flight from Dallas to Montreal, a fire broke out near the rear toilet, which engulfed the whole airplane in toxic smoke and fumes. The flight immediately diverted to Cincinnati airport. I later learned that the fellow I had tutored to be my replacement on the MSR project, as well as my former boss—a VP in Leigh's sales and marketing department, and a great guy who also had become a good friend—both died of smoke inhalation, along with twenty-one other passengers. I couldn't help thinking what might have been.

Philip Moylan

It was Philip Moylan's employer who suggested that he return to obtain a Bachelor of Engineering degree on a part-time basis.

Six years later, he is about to become the first student to graduate with an engineering degree earned as a part-time student.

Philip admitted that the past six years have been gruelling, but he said that the opportunity of mixing practical work experience with theoretical study was particularly rewarding.

Currently chief design engineer with H. Aass Aero Engineering Limited, Philip's job is to predetermine stress that occurs in aeroplanes structures that have been modified to carry equipment such as infra-red cameras or insecticide sprayers. Such modifications can cause vibrations so intense that they can literally tear a wing from a plane, he said.

As a student at Carleton, Philip "challenged" two courses required for his degree. University rules enable students with the necessary qualifications to count their expertise for credit.

Philip was able to use his work experience to get credit for a drafting course and a third-year design studies course.

Instead of taking the design course, he prepared two case studies. One was a general analysis of the effects that the suspension of a crop-spraying rig from a wing would have on aircraft characteristics and handling. This study, based on his work, will be used in the curriculum of the design course in the future.

Philip also has a patent pending on the design of a hand-held underwater power drill which he and another student, David Harwood, completed for a fourth-year design course.

Although he is considering working toward his master's degree, Philip said he will do it at a much slower pace.

Right now, he has his sights set on obtaining his licence to become a design approval representative. With this licence, he will be qualified to inspect modified planes and authorize them fo fly. At present, he said, there are only 15 such representatives in Canada.

Carleton University graduation news

Slide from my Leigh Canada House presentation—crash recorder on a Beech 18

Deployable crash recorder mounted on the tail of a Beech 18 aircraft

A "flyable" crash recorder on a Boeing chopper, designed to be ejected during a crash

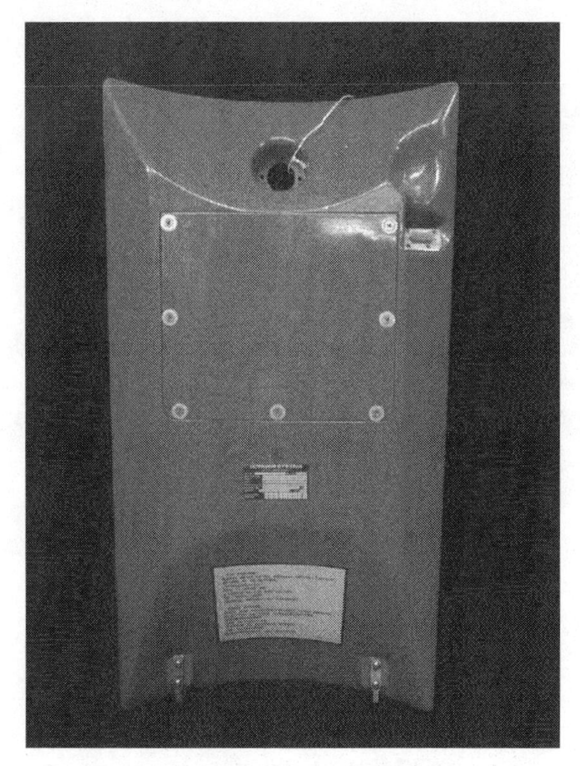

A Leigh crash recorder

United States Patent [19] [11] **3,906,511**

Kelemen [45] **Sept. 16, 1975**

[54] **CASSETTE SCRATCH STRAIN RECORDER**

[75] Inventor: **Jozsef W. Kelemen,** Ottawa, Canada

[73] Assignee: **Leigh Instruments Limited,** Carleton Place, Canada

[22] Filed: **Nov. 19, 1973**

[21] Appl. No.: **416,958**

[52] **U.S. Cl.** .... **346/7;** 74/88; 192/41 S; 346/136

[51] **Int. Cl.²** **G01B 5/30;** G01D 9/38

[58] **Field of Search** .... 346/7, 77, 136, 145; 73/88; 33/147 D, 148 D; 74/88; 192/12 BA, 41 S

[56] **References Cited**

UNITED STATES PATENTS

3,389,611	6/1968	Bey	74/88
3,754,276	8/1973	Endres	346/7
3,825,934	7/1974	Price et al.	346/7

Primary Examiner—Joseph W. Hartary

[57] **ABSTRACT**

A recording instrument is disclosed for measuring and recording changes in linear distance between two predetermined points. The instrument is comprised of a recording tape and a scribing means for marking the changes in linear distance on the tape. The scribing means moves in a direction which is transverse to the longitudinal axis of the tape. The instrument includes a tape housing means which accommodates the tape so that the tape is free to move along its longitudinal axis but is substantially constrained from transverse movement. The instrument additionally includes a tape drive means for longitudinally advancing the tape under the influence of both positive and negative changes in the linear distance. The tape drive means includes a linear motion amplifier and a reciprocating to rotary motion converter.

10 Claims, 5 Drawing Figures

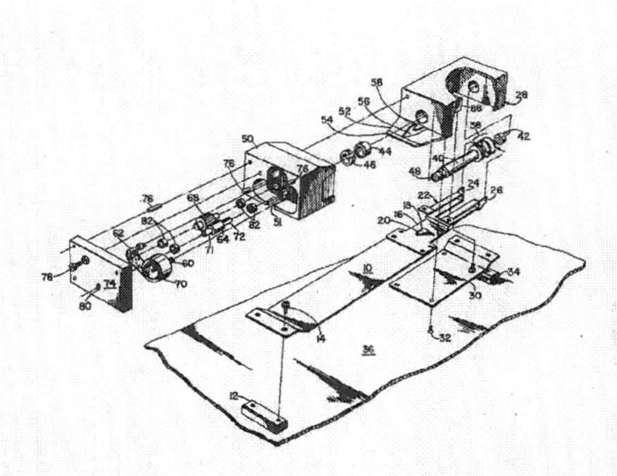

US Patent for Leigh Instruments Mechanical Strain Recorder (MSR)

INTO THE FUTURE WITH AIR CANADA

AIR CANADA

LEIGH INSTRUMENTS LIMITED

is proud to have been selected by

AIR CANADA

to supply

FLIGHT DATA RECORDER SYSTEMS

for its fleet of DC 8 and DC 9 Aircraft

To assist in maintaining its remarkable record of efficiency, safety and dependability, Air Canada has chosen Leigh Instruments' light weight, all solid state equipment to provide a most reliable means of recording and preserving aircraft flight data.

LEIGH
INSTRUMENTS
LIMITED
CARLETON PLACE, ONTARIO, CANADA

Playing guitar with my band, Alliance, at a Leigh Instruments Christmas party

Acknowledgements

Dedications

Early in the new century, Dick Hansen and I were in Long Beach, California, at an aviation conference. We were visiting the Queen Mary during some time off, when we got a call from one of our former associates at Flight Dynamics who gave us the news that Jim Gooden had passed away. It was very upsetting for us both, and for me personally. Jim had been a great mentor to me in my career selling frozen turkey heads. We had stayed in touch after he retired, as he had wanted to keep abreast of all the exciting happenings at the company. I would miss our chats terribly.

A few years later, in 2004, Lyn Sorensen, who was now working at BBJ in contracts, phoned to tell me that Borge Boeskov had also passed away, at age 67. Another shocking blow. Borge had become a great friend and a terrific visionary, both at Boeing and later, as the president of BBJ. I would miss him also. When my wife and I were invited to the first-ever hospitality suite hosted by BBJ at the NBAA in Orlando, Florida, I went to get some drinks from the bar, leaving Chris alone. Borge saw her standing there, and left the people he was talking with to go over and say hi and welcome her. In a condolence letter I wrote to his wife, Sandy, I told her that "the true measure of a man is how he treats those who can do nothing for him." Borge was a prince. I will never forget him.

In 2007, I got a call from Dean Schwab to tell me that Bruce Kennedy, the original president "across the street" at Alaska Airlines, and the one who had formally approved our very first HGS purchase there, had been killed in an aircraft accident. He had been flying his own Cessna in Central Washington when the aircraft went down. Another blow.

This book is dedicated to Jim, Borge, Bruce, and other aviation industry visionaries who have left us. Without them, our HGS might never have reached the successes that it did. These individuals played a huge and key role in the development of the commercial HUD marketplace and specifically in helping get our HGS to where it is today. I recognize them here for their vision, insights, and professional efforts to bring a new level of enhanced safety and improvement in operations to the air transport industry:

Jim Gooden, Director Sales & Marketing, Flight Dynamics
Berk Greene, AFS 400, Federal Aviation Administration
Borge Boeskov, President, Boeing Business Jets
Bruce Kennedy, President, Alaska Airlines
Jan Strom, VP Flight Ops, SAS

Recognition

The HUD effort also would not have been possible without the dedication and support of many key folks at the OEMs, at the airlines and, of course, at Flight Dynamics, who pushed the outer boundaries of our industry and challenged the status quo. These are some of them, in the order that I met them, with their positions at that time:

Mike Gleason, Flight Operations, Flight Dynamics
John Desmond, President, Flight Dynamics
Bob Wood, Manager Optics Engineering, Flight Dynamics
Doug Ford, Principal Engineer Flight Controls, Flight Dynamics
Ken Zimmerman, Manager Software Engineering, Flight Dynamics
Norman Jee, Manager Mechanical Engineering, Flight Dynamics
Randy Foster, Product Support Flight Dynamics
Dick Hansen, Director HGS Flight Ops, Flight Dynamics
Tom Geiger, Marketing Analyst, Flight Dynamics
Tom Johnson, Assistant Chief Pilot, Alaska Airlines
Dean Schwab, Flight Ops Technical, Alaska Airlines
Larry Hecker, VP Flight Ops Western Airlines
Chris Longridge, VP Marketing, Boeing Company
Carl Lund, B737 Development Engineer, Boeing Company
Jim Von Der Linn, Regional Director/737 Model Manager, Boeing Company
Lyn Sorenson, Manager HUD Procurement, Boeing Company
George Kanellis, Marketing Analyst, Flight Dynamics
Tom Kilbane, Marketing Manager Europe, Flight Dynamics
Troy Menken, Flight Ops Technical, Morris Air
Paul Sterbenz, VP Flight Ops, Southwest Airlines
Joe Marott, Director Training, Southwest Airlines
Bob George, Customer Engineer Southwest, Boeing Company
Carter Chapman, Chief Technical Pilot, Delta Air Lines
BJ Smith, Technical Pilot, Delta Airlines
Greg Saylor, Technical pilot, Delta Airlines

It is indeed an honor and a privilege to have worked with each one of you, and to have gotten to know you as well as I did.

Thanks

Thanks to the good folks who helped me with this book by sharing their great memories (much better than mine) and their stories, corrections, and pictures. Getting input and encouragement from friends in the industry brought back many wonderful memories for me, and made this book a joy to work on.

And thanks to my editor, Karen Story, for a great job keeping my grammar on track and polishing my story for such a professional reading audience.

From the bottom of my heart to all of you…THANK YOU!

Made in the USA
Las Vegas, NV
08 August 2022